WORKSHEETS
WITH THE MATH COACH

GEX, INCORPORATED

BEGINNING &
INTERMEDIATE ALGEBRA
FOURTH EDITION

John Tobey
North Shore Community College

Jeffrey Slater
North Shore Community College

Jamie Blair
Orange Coast College

Jennifer Crawford
Normandale Community College

PEARSON

Boston Columbus Indianapolis New York San Francisco Upper Saddle River
Amsterdam Cape Town Dubai London Madrid Milan Munich Paris Montreal Toronto
Delhi Mexico City Sao Paulo Sydney Hong Kong Seoul Singapore Taipei Tokyo

The author and publisher of this book have used their best efforts in preparing this book. These efforts include the development, research, and testing of the theories and programs to determine their effectiveness. The author and publisher make no warranty of any kind, expressed or implied, with regard to these programs or the documentation contained in this book. The author and publisher shall not be liable in any event for incidental or consequential damages in connection with, or arising out of, the furnishing, performance, or use of these programs.

Reproduced by Pearson from electronic files supplied by the author.

ISBN-13: 978-0-321-78055-3
ISBN-10: 0-321-78055-8

4 5 6 EBM 16 15 14 13 12

www.pearsonhighered.com

Worksheets with the Math Coach

Beginning and Intermediate Algebra, Fourth Edition

Table of Contents

Chapter 0 Prealgebra Review
0.1 Simplifying Fractions

Vocabulary
whole numbers • fractions • numerator • denominator • numerals
simplest form • reduced form • simplifying • reducing • natural numbers
counting numbers • factor • prime numbers • lowest terms • multiplicative identity
proper fraction • improper fraction • mixed number

1. The whole numbers 1, 2, 3, 4, 5, 6, 7, 8, and 9 are called _____ or counting numbers.

2. In the fraction $\dfrac{1}{25}$, 1 is the numerator and 25 is the _____.

3. When we obtain an equivalent fraction in simplest form (or reduced form), the new simplified fraction is said to be in _____.

4. The natural number factors of _____ are 1 and themselves.

Example	**Student Practice**
1. Simplify each fraction.	**2.** Simplify each fraction.
(a) $\dfrac{14}{21}$	**(a)** $\dfrac{33}{39}$
First factor 14 and 21, $\dfrac{14}{21} = \dfrac{7 \times 2}{7 \times 3}$.	
Now divide the numerator and denominator by 7.	
$\dfrac{14}{21} = \dfrac{\cancel{7} \times 2}{\cancel{7} \times 3} = \dfrac{2}{3}$	
(b) $\dfrac{20}{70}$	**(b)** $\dfrac{126}{210}$
Factor 20 and 70, $\dfrac{20}{70} = \dfrac{2 \times 2 \times 5}{7 \times 2 \times 5}$.	
Now divide the numerator and denominator by both 2 and 5.	
$\dfrac{20}{70} = \dfrac{2 \times \cancel{2} \times \cancel{5}}{7 \times \cancel{2} \times \cancel{5}} = \dfrac{2}{7}$	

Vocabulary Answers: 1. natural numbers 2. denominator 3. lowest terms 4. prime numbers

Example	Student Practice
3. Simplify the fraction $\dfrac{7}{21}$. First factor 7 and 21, $\dfrac{7}{21} = \dfrac{7 \times 1}{7 \times 3}$. Now divide the numerator and denominator by 7. $\dfrac{7}{21} = \dfrac{\cancel{7} \times 1}{\cancel{7} \times 3} = \dfrac{1}{3}$ Notice that all the prime numbers in the numerator divided out. When this happens, we must remember that 1 is left in the numerator.	**4.** Simplify the fraction $\dfrac{13}{169}$.
5. Simplify the fraction $\dfrac{70}{10}$. $\dfrac{70}{10} = \dfrac{7 \times \cancel{5} \times \cancel{2}}{\cancel{5} \times \cancel{2} \times 1} = 7$ Notice that all the prime numbers in the denominator divided out. When this happens, we do not need to leave 1 in the denominator because the answer is a whole number.	**6.** Simplify the fraction $\dfrac{423}{47}$.
7. Cindy got 48 out of 56 questions correct on a test. Write this as a fraction in simplest form. Express as a fraction the number of correct responses out of the total number of questions on the test. 48 out of 56 $\rightarrow \dfrac{48}{56}$ Express this fraction in simplest form. Factor 48 and 56, then divide out common factors. $\dfrac{48}{56} = \dfrac{6 \times \cancel{8}}{7 \times \cancel{8}} = \dfrac{6}{7}$ Cindy answers the questions correctly $\dfrac{6}{7}$ of the time.	**8.** Last June, Jacob went to the gym 18 out of the 30 days. Write this as a fraction in simplest form.

2

Example	Student Practice
9. Change $\dfrac{7}{4}$ to a mixed number or to a whole number.	**10.** Change $\dfrac{63}{7}$ to a mixed number or to a whole number.

Divide the denominator into the numerator.

$$\frac{7}{4} = 7 \div 4$$

$$\begin{array}{r} 1 \\ 4\overline{)7} \\ \underline{4} \\ 3 \end{array} \quad \text{Remainder}$$

The quotient, 1, is the whole-number part of the mixed number. The remainder from the division, 3, is the numerator of the fraction. The denominator of the fraction remains unchanged.

Thus, $\dfrac{7}{4} = 1\dfrac{3}{4}$.

11. Change $3\dfrac{1}{7}$ to an improper fraction.	**12.** Change $6\dfrac{5}{12}$ to an improper fraction

To change a mixed number to an improper fraction, multiply the whole number by the denominator. Add this to the numerator. The result is the new numerator. The denominator does not change.

$$3\frac{1}{7} = \frac{(3 \times 7) + 1}{7} = \frac{21 + 1}{7} = \frac{22}{7}$$

Thus, $3\dfrac{1}{7} = \dfrac{22}{7}$.

Example	Student Practice
13. Find the missing number.	**14.** Find the missing number.
(a) $\dfrac{3}{5} = \dfrac{?}{25}$	(a) $\dfrac{1}{9} = \dfrac{?}{99}$

A fraction can be changed to an equivalent fraction with a different denominator by multiplying both numerator and denominator by the same number. Observe that we need to multiply the denominator by 5 to obtain 25. So we multiply the numerator 3 by 5 also.

$$\dfrac{3 \times 5}{5 \times 5} = \dfrac{15}{25}$$

The desired number is 15.

(b) $\dfrac{2}{9} = \dfrac{?}{36}$

(b) $\dfrac{4}{7} = \dfrac{?}{35}$

Observe that $9 \times 4 = 36$. We need to multiply the numerator by 4 to get the new numerator.

$$\dfrac{2 \times 4}{9 \times 4} = \dfrac{8}{36}$$

The desired number is 8.

Extra Practice

1. Simplify the fraction $\dfrac{60}{15}$.

2. Change $\dfrac{556}{10}$ to a mixed number.

3. Change $1\dfrac{12}{17}$ to an improper fraction.

4. Find the missing numerator, $\dfrac{8}{15} = \dfrac{?}{120}$.

Concept Check

Explain in your own words how to change a mixed number to an improper fraction.

Chapter 0 Prealgebra Review
0.2 Adding and Subtracting Fractions

Vocabulary

fraction   •   numerator   •   denominator   •   common denominator
least common denominator   •   prime numbers   •   mixed numbers   •   perimeter

1. The _____ of two or more fractions is the smallest whole number that is exactly divisible by each denominator of the fractions.

2. Before you can add or subtract fractions, they must have the same _____.

3. The distance around a polygon is called the _____.

4. When adding or subtracting _____ first change them to improper fractions.

Example	**Student Practice**
1. Add the fractions. Simplify your answer whenever possible.	2. Add the fractions. Simplify your answer whenever possible.
(a) $\dfrac{5}{7}+\dfrac{1}{7}$	(a) $\dfrac{1}{4}+\dfrac{3}{4}$
Add the numerators and keep the denominator the same. $$\dfrac{5}{7}+\dfrac{1}{7}=\dfrac{5+1}{7}=\dfrac{6}{7}$$ This fraction cannot be simplified further.	
(b) $\dfrac{1}{8}+\dfrac{3}{8}+\dfrac{2}{8}$	(b) $\dfrac{7}{10}+\dfrac{5}{10}+\dfrac{3}{10}$
Add the numerators and keep the denominator the same. $$\dfrac{1}{8}+\dfrac{3}{8}+\dfrac{2}{8}=\dfrac{1+3+2}{8}=\dfrac{6}{8}$$ Now simplify. $$\dfrac{6}{8}=\dfrac{3}{4}$$	

Vocabulary Answers: 1. least common denominator 2. denominator 3. perimeter 4. mixed numbers

Example	Student Practice
3. Subtract the fractions $\dfrac{9}{11} - \dfrac{2}{11}$. Simplify your answer if possible. Subtract the numerators and keep the denominator the same. $$\dfrac{9}{11} - \dfrac{2}{11} = \dfrac{9-2}{11} = \dfrac{7}{11}$$ This fraction cannot be simplified further.	**4.** Subtract the fractions $\dfrac{7}{8} - \dfrac{5}{8}$. Simplify your answer if possible.
5. Find the LCD of $\dfrac{5}{6}$ and $\dfrac{1}{15}$ using the prime factor method. Write each denominator as the product of prime factors. The LCD is a product containing each different prime factor. $$6 = 2 \cdot 3$$ $$15 = \downarrow \ 3 \cdot 5$$ $$\ \ \ \ \ \downarrow \ \downarrow \ \downarrow$$ $$\text{LCD} = 2 \cdot 3 \cdot 5$$ The different factors are 2, 3, and 5, and each factor appears at most once in any one denominator. Multiply the factors to find the LCD. $$\text{LCD} = 2 \cdot 3 \cdot 5 = 30$$	**6.** Find the LCD of $\dfrac{1}{10}$ and $\dfrac{6}{35}$ using the prime factor method.
7. Find the LCD of $\dfrac{5}{12}$, $\dfrac{1}{15}$, and $\dfrac{7}{30}$. Write each denominator as the product of prime factors. The LCD is a product containing each different factor, with the factor 2 appearing twice since it occurs twice in the factorization of 12. $$12 = 2 \cdot 2 \cdot 3$$ $$15 = \downarrow \ \ \ \ 3 \cdot 5$$ $$\ \ \ \ \ \downarrow$$ $$30 = \downarrow \ 2 \cdot 3 \cdot 5$$ $$\ \ \ \ \ \downarrow \ \downarrow \ \downarrow \ \downarrow$$ $$\text{LCD} = 2 \cdot 2 \cdot 3 \cdot 5 = 60$$	**8.** Find the LCD of $\dfrac{11}{18}$, $\dfrac{2}{21}$, and $\dfrac{13}{42}$.

6

Example	Student Practice

9. Combine. $\dfrac{1}{5}+\dfrac{1}{6}-\dfrac{3}{10}$

First find the LCD.

$$5 = 5$$
$$6 = 2\cdot3$$
$$10 = 5\cdot2 \downarrow$$
$$\downarrow\ \downarrow\ \downarrow$$
$$LCD = 5\cdot2\cdot3 = 30$$

Now we change $\dfrac{1}{5}$, $\dfrac{1}{6}$, and $\dfrac{3}{10}$ to equivalent fractions that have the LCD for a denominator.

$$\frac{1}{5}=\frac{?}{30}\qquad \frac{1\times6}{5\times6}=\frac{6}{30}$$

$$\frac{1}{6}=\frac{?}{30}\qquad \frac{1\times5}{6\times5}=\frac{5}{30}$$

$$\frac{3}{10}=\frac{?}{30}\qquad \frac{3\times3}{10\times3}=\frac{9}{30}$$

Combine the three fractions.

$$\frac{1}{5}+\frac{1}{6}-\frac{3}{10}=\frac{6}{30}+\frac{5}{30}-\frac{9}{30}$$
$$=\frac{6+5-9}{30}$$
$$=\frac{2}{30}$$

Now simplify.

$$\frac{2}{30}=\frac{1}{15}$$

Thus, $\dfrac{1}{5}+\dfrac{1}{6}-\dfrac{3}{10}=\dfrac{1}{15}$.

10. Combine. $\dfrac{1}{4}+\dfrac{13}{18}-\dfrac{5}{54}$

7

Example	Student Practice

11. Manuel is enclosing a triangular shaped exercise yard for his new dog. He wants to determine how many feet of fencing he will need. The sides of the yard measure $20\frac{3}{4}$ feet, $15\frac{1}{2}$ feet, and $18\frac{1}{8}$ feet. What is the perimeter of (the total distance around) the triangle?

Begin by drawing a picture. Find the perimeter by adding up the lengths of the three sides of the triangle. To add mixed numbers, change them to improper fractions then add.

$$20\frac{3}{4}+15\frac{1}{2}+18\frac{1}{8} = \frac{83}{4}+\frac{31}{2}+\frac{145}{8}$$

$$= \frac{166}{8}+\frac{124}{8}+\frac{145}{8}$$

$$= \frac{435}{8} = 54\frac{3}{8}$$

He will need $54\frac{3}{8}$ feet of fencing.

12. Find the perimeter of the parallelogram shaped park below.

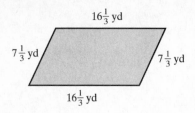

$16\frac{1}{3}$ yd

$7\frac{1}{3}$ yd $7\frac{1}{3}$ yd

$16\frac{1}{3}$ yd

Extra Practice

1. Find the LCD of $\frac{6}{25}$ and $\frac{99}{100}$. Do not combine the fractions; only find the LCD.

2. Combine $\frac{31}{45}-\frac{2}{15}$. Be sure to simplify your answer if possible.

3. Combine $\frac{7}{9}+2\frac{5}{12}$. Be sure to simplify your answer if possible.

4. Combine $6\frac{3}{11}-2\frac{17}{22}$. Be sure to simplify your answer if possible.

Concept Check

Explain how you would find the LCD of the fractions $\frac{4}{21}$ and $\frac{5}{18}$.

Name: _____   Date: _____

Instructor: _____   Section: _____

Chapter 0 Prealgebra Review
0.3 Multiplying and Dividing Fractions

Vocabulary

fractions   •   numerators   •   denominators   •   invert and multiply method
common factors   •   prime numbers   •   mixed numbers   •   complex fraction

1. A(n) _____ has one fraction in the numerator and one fraction in the denominator.

2. To multiply two fractions, multiply the _____ and multiply the denominators.

3. Use the _____ to divide two fractions.

4. When multiplying or dividing _____, first change them to improper fractions.

Example	**Student Practice**
1. Multiply. (a) $\dfrac{3}{5} \times \dfrac{5}{7}$	**2.** Multiply. (a) $\dfrac{2}{7} \times \dfrac{7}{17}$

Example (continued)

Multiply the numerators and multiply the denominators.

$$\frac{3}{5} \times \frac{5}{7} = \frac{3 \cdot 5}{5 \cdot 7}$$

Divide the numerator and denominator by 5 and simplify.

$$\frac{3}{5} \times \frac{5}{7} = \frac{3 \cdot 5}{5 \cdot 7} = \frac{3 \cdot \overset{1}{\cancel{5}}}{\underset{1}{\cancel{5}} \cdot 7} = \frac{3}{7}$$

(b) $\dfrac{15}{8} \times \dfrac{10}{27}$

(b) $\dfrac{55}{12} \times \dfrac{4}{45}$

If we factor each number, we can see the common factors. Remove common factors and multiply.

$$\frac{15}{8} \times \frac{10}{27} = \frac{\overset{1}{\cancel{3}} \cdot 5}{2 \cdot 2 \cdot \underset{1}{\cancel{2}}} \times \frac{5 \cdot \overset{1}{\cancel{2}}}{\underset{1}{\cancel{3}} \cdot 3 \cdot 3} = \frac{25}{36}$$

Vocabulary Answers: 1. complex fraction 2. numerators 3. invert and multiply method 4. mixed numbers

Example	Student Practice

3. Multiply $7 \times \dfrac{3}{5}$.

Write the whole number as a fraction whose denominator is 1.

$$7 \times \frac{3}{5} = \frac{7}{1} \times \frac{3}{5}$$

Follow the multiplication rule for fractions. Multiply numerators and multiply denominators.

$$7 \times \frac{3}{5} = \frac{7}{1} \times \frac{3}{5} = \frac{21}{5} \text{ or } 4\frac{1}{5}$$

4. Multiply $25 \times \dfrac{4}{5}$.

---

5. How do we find the area of a rectangular field $3\dfrac{1}{3}$ miles long and $2\dfrac{1}{2}$ miles wide?

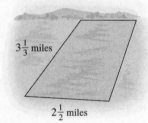

$3\frac{1}{3}$ miles

$2\frac{1}{2}$ miles

To find the area, we multiply length times width, $3\dfrac{1}{3} \times 2\dfrac{1}{2}$.

To multiply two mixed numbers, change them to improper fractions then follow the multiplication rule for fractions.

$$3\frac{1}{3} \times 2\frac{1}{2} = \frac{10}{3} \times \frac{5}{2} = \frac{\cancel{2} \cdot 5}{3} \times \frac{5}{\cancel{2}} = \frac{25}{3}$$

Write the answer as a mixed number.
The area is $8\dfrac{1}{3}$ square miles.

6. Find the area of a rectangular pool that is $7\dfrac{1}{12}$ meters long and $6\dfrac{3}{10}$ meters wide.

Example	Student Practice
7. Divide $\dfrac{2}{5} \div \dfrac{3}{10}$.	**8.** Divide $\dfrac{1}{9} \div \dfrac{5}{6}$.

To divide two fractions, invert the second fraction (the divisor) and multiply.

$$\frac{2}{5} \div \frac{3}{10} = \frac{2}{5} \times \frac{10}{3}$$

Remove common factors and simplify.

$$\frac{2}{5} \div \frac{3}{10} = \frac{2}{5} \times \frac{10}{3}$$

$$= \frac{2}{\cancel{5}} \times \frac{\cancel{5} \cdot 2}{3} = \frac{4}{3} \text{ or } 1\frac{1}{3}$$

9. Divide $\dfrac{1}{3} \div 2$.	**10.** Divide $7 \div \dfrac{1}{5}$.

Write the whole number as a fraction. Then invert the second fraction (the divisor) and multiply.

$$\frac{1}{3} \div 2 = \frac{1}{3} \div \frac{2}{1} = \frac{1}{3} \times \frac{1}{2} = \frac{1}{6}$$

11. Divide $\dfrac{\dfrac{3}{7}}{\dfrac{3}{5}}$.	**12.** Divide $\dfrac{\dfrac{7}{20}}{\dfrac{14}{15}}$.

This division is in the form of a complex fraction. Write this problem as a standard division first, then complete the problem using the rule for division.

$$\frac{\dfrac{3}{7}}{\dfrac{3}{5}} = \frac{3}{7} \div \frac{3}{5} = \frac{\cancel{3}}{7} \times \frac{5}{\cancel{3}} = \frac{5}{7}$$

Example	Student Practice

13. Divide $2\dfrac{1}{3} \div 3\dfrac{2}{3}$.

To divide mixed numbers, change the mixed numbers to improper fractions and then use the rule for dividing fractions.

$$2\dfrac{1}{3} \div 3\dfrac{2}{3} = \dfrac{7}{3} \div \dfrac{11}{3} = \dfrac{7}{\cancel{3}} \times \dfrac{\cancel{3}}{11} = \dfrac{7}{11}$$

14. Divide $\dfrac{10}{2\dfrac{4}{5}}$.

---

15. A chemist has 96 fluid ounces of a solution. She pours the solution into test tubes. Each test tube holds $\dfrac{3}{4}$ fluid ounces. How many test tubes can she fill?

Divide the total number of ounces by the number of ounces in each test tube.

$$96 \div \dfrac{3}{4} = \dfrac{96}{1} \div \dfrac{3}{4}$$

$$= \dfrac{96}{1} \times \dfrac{4}{3} = \dfrac{\cancel{3} \cdot 32}{1} \times \dfrac{4}{\cancel{3}} = 128$$

She will be able to fill 128 test tubes.

16. A graphic designer created 5 images for the evening news program in $6\dfrac{1}{4}$ hours. How many images did the designer create per hour?

Extra Practice

1. Multiply $\dfrac{3}{7} \times \dfrac{5}{12} \times \dfrac{8}{35}$. Simplify your answer if possible.

2. Multiply $7\dfrac{1}{2} \times 2\dfrac{2}{5}$. Simplify your answer if possible.

3. Divide $\dfrac{2}{3} \div \dfrac{4}{9}$. Simplify your answer if possible.

4. Mario cuts $59\dfrac{1}{2}$ inches of string into pieces that are exactly $4\dfrac{1}{4}$ inches long. How many pieces of string does Mario cut?

Concept Check

Explain the steps you would take to perform the calculation $3\dfrac{1}{4} \div 2\dfrac{1}{2}$.

Chapter 0 Prealgebra Review
0.4 Using Decimals

Vocabulary

decimal • decimal points • decimal places • terminal zeros • divisor
dividend • quotient • caret • fraction

1. A fraction whose denominator is 10, 100, and so on can be expressed as a _____.

2. Divide the dividend by the _____.

3. The number of digits to the right of the decimal point is also the number of _____.

4. To add or subtract decimals, write in column form and line up the _____.

Example	Student Practice
1. Write each of the following decimals as a fraction or mixed number. State the number of decimal places. Write out in words the way the number would be spoken. **(a)** 0.6 In fraction form, $0.6 = \dfrac{6}{10}$. 0.6 has 1 decimal place because it has 1 digit to the right of the decimal point. Use a place value chart if necessary to determine the word name. Written out, 0.6 is six-tenths. **(b)** 1.38 In fraction form, $1.38 = 1\dfrac{38}{100}$. 1.38 has 2 decimal places because it has 2 digits to the right of the decimal point. Use a place value chart if necessary to determine the word name. Written out, 1.38 is one and thirty-eight hundredths.	**2.** Write each of the following decimals as a fraction or mixed number. State the number of decimal places. Write out in words the way the number would be spoken. **(a)** 0.00009 **(b)** 4.025

Vocabulary Answers: 1. decimal 2. divisor 3. decimal places 4. decimal points

Example	Student Practice
3. Write $\dfrac{2}{11}$ as a decimal.	**4.** Write $\dfrac{13}{15}$ as a decimal.

$$\begin{array}{r} 0.1818 \\ 11\overline{)2.0000} \\ \underline{1\ 1} \\ 90 \\ \underline{88} \\ 20 \\ \underline{11} \\ 90 \\ \underline{88} \\ 2 \end{array}$$

Thus, $\dfrac{2}{11} = 0.1818\ldots = 0.\overline{18}$.

5. Write 0.138 as a fraction and simplify if possible.	**6.** Write 0.72 as a fraction and simplify if possible.

To convert from a decimal to a fraction, write the decimal as a fraction with a denominator of 10, 100, 1000, and so on, and simplify the result.

Write 0.138 as a fraction with numerator 138 and denominator 1000. Then simplify.

$$0.138 = \frac{138}{1000} = \frac{69}{500}$$

7. Subtract $127.32 - 38.48$.	**8.** Add $21.16 + 2.04 + 13.91$.

To add or subtract decimals, write in column form and line up the decimal points. Then add or subtract the digits.

$$\begin{array}{r} 127.32 \\ -\ \ 38.48 \\ \hline 88.84 \end{array}$$

14

Example	Student Practice
9. Multiply 2.56×0.003.	**10.** Multiply 0.85×0.0005.

9. Multiply 2.56×0.003.

To multiply decimals, first multiply as with whole numbers. To determine the position of the decimal point, count the total number of decimal places in the two numbers being multiplied. This will determine the number of decimal places that should appear in the answer.

$$\begin{array}{r} 2.56 \quad \text{(two decimal places)} \\ \times \quad 0.003 \quad \text{(three decimal places)} \\ \hline 0.00768 \quad \text{(five decimal places)} \end{array}$$

Thus, $2.56 \times 0.003 = 0.00768$.

---

11. Divide $16.2 \div 0.027$.

There are three decimal places in the divisor, so we move the decimal point three places to the right in the divisor and dividend and mark the new position by a caret.

Note that we must add two zeros to 16.2 in order to do this.

$$0.027_{\wedge} \overline{)16.200_{\wedge}} \quad \overset{600.}{}$$

Now perform the division as with whole numbers. The decimal point in the answer is directly above the caret.

$$0.027_{\wedge} \overline{)16.200_{\wedge}} \quad \overset{600.}{}$$
$$\underline{16\ 2}$$
$$000$$

Thus, $16.2 \div 0.027 = 600$.

12. Divide $25.2 \div 0.0003$.

Example	Student Practice
13. Multiply 0.0026×1000. When multiplying by 10, 100, 1000, and so on, for every zero in the multiplier, move the decimal point one place to the right. Since 1000 has 3 zeros, move the decimal point 3 places to the right. $0.0026 \times 1000 = 2.6$	**14.** Multiply 0.93×100.
15. Divide $0.0038 \div 100$. When dividing by 10, 100, 1000, and so on, for every zero in the divisor, move the decimal point one place to the left. Since 100 has 2 zeros, move the decimal point 2 places to the left. $0.0038 \div 100 = 0.000038$	**16.** Divide $7138.3 \div 10$.

Extra Practice

1. Write 32.082 as a fraction in simplified from. Write the value in words.

2. Subtract $14.03 - 7.8932$.

3. Divide $1.1592 \div 0.06$.

4. Multiply 16.785×100 by moving the decimal point.

Concept Check

Explain how you would place the decimal points when performing the calculation $0.252 \div 0.0035$.

Chapter 0 Prealgebra Review
0.5 Percents, Rounding, and Estimating

Vocabulary

decimal   •   decimal point   •   percent   •   % symbol   •   estimation   •   nonzero digit

1. A _____ is a fraction that has a denominator of 100.

2. To estimate by rounding, first round each number so that there is one _____.

3. _____ is the process of finding an approximate answer.

4. To change a percent to a decimal, move the _____ two places to the left and remove the % symbol.

Example	Student Practice
1. Change to a percent. **(a)** 0.0364 We move the decimal point two places to the right and add the % symbol. 0.0364 = 3.64% **(b)** 0.4 0.4 = 40%	**2.** Change to a percent. **(a)** 0.00019 **(b)** 0.6
3. Change to a percent. **(a)** 2.938 We move the decimal point two places to the right and add the % symbol. 2.938 = 293.8% **(b)** 4.5 4.5 = 450%	**4.** Change to a percent. **(a)** 5.95 **(b)** 7.9

Vocabulary Answers: 1. percent 2. nonzero digit 3. estimation 4. decimal point

Example	Student Practice
5. Change to a decimal. **(a)** 4% First we move the decimal point two places to the left. Then we remove the % symbol. $4\% = 4\underset{\uparrow}{.}\% = 0.04$ The unwritten decimal point is understood to be here. **(b)** 254.8% $254.8\% = 2.548$	**6.** Change to a decimal. **(a)** 0.8% **(b)** 101.2%
7. Find 182% of 12. To find the percent of a number, change the percent to a decimal and multiply the number by the decimal. 182% of $12 = 1.82 \times 12 = 21.84$	**8.** Find 5% of 639.
9. A store is having a sale of 35% off the retail price of all sofas. Melissa wants to buy a particular sofa that normally sells for $595. **(a)** How much will Melissa save if she buys the sofa on sale? Find 35% of $595. $0.35 \times 595 = 208.25$ Thus, Melissa will save $208.25. **(b)** What will the purchase price be if Melissa buys the sofa on sale? The purchase price is the difference between the original price and the amount saved. $595.00 - 208.25 = 386.75$ If Melissa buys the sofa on sale, she will pay $386.75.	**10.** A store is having a sale of 65% off the retail price of all boots. Marcus wants to buy a particular pair of boots that normally sells for $123. **(a)** How much will Marcus save if he buys the boots on sale? **(b)** What will the purchase price be if Marcus buys the boots on sale?

Example	Student Practice
11. What percent of 24 is 15? First, write a fraction with the two numbers. The number after the word of is always the denominator, and the other number is the numerator. $$\frac{15}{24}$$ Simplify the fraction. $$\frac{15}{24} = \frac{5}{8}$$ Change the simplified fraction to a decimal. $$\frac{5}{8} = 0.625$$ Express the decimal as a percent. $$0.625 = 62.5\%$$ Thus, 62.5% of 24 is 15.	**12.** What percent of 72 is 63?
13. Marcia made 29 shots on goal during the last high school field hockey game. She actually scored a goal 8 times. What percent of her total shots were goals? Round your answer to the nearest whole percent. We want to know what percent of 29 is 8. This relationship expressed as a fraction is $\frac{8}{29}$. This fraction cannot be reduced. The next step is to write the fraction as a decimal. $$\frac{8}{29} = 0.2758...$$ Change the decimal to a percent and round to the nearest whole number. $$0.2758... = 27.58...\% \approx 28\%$$ Thus, Marcia scored a goal about 28% of the time she made a shot on goal.	**14.** In a shipment of 600 widgets, 38 widgets are warped. What percent of the total widgets are warped? Round your answer to the nearest whole percent.

Example	Student Practice

15. The four walls of a college classroom are $22\frac{1}{4}$ feet long and $8\frac{3}{4}$ feet high. A painter needs to know the area of these four walls in square feet. Since paint is sold in gallons, an estimate will do. Use estimation by rounding to approximate the area of the four walls.

To estimate by rounding, round each number so that there is one nonzero digit. Round $22\frac{1}{4}$ feet to 20 feet.

Round $8\frac{3}{4}$ feet to 9 feet.

Multiply 20×9 to obtain an estimate of the area of one wall. Multiply $20 \times 9 \times 4$ to obtain an estimate of the area of all four walls.

$20 \times 9 \times 4 = 720$

Our estimate for the painter is 720 square feet of wall space.

16. A landscaper is to seed a lawn that is $47\frac{1}{2}$ meters long and $18\frac{1}{4}$ meters wide. The landscaper needs to know the area of the lawn in square meters to determine how much grass seed to buy. Use estimation by rounding to approximate the area of the lawn.

Extra Practice

1. Change 0.00576 to a percent.

2. Change 100% to a decimal.

3. 65 is what percent of 25?

4. Follow the principles of estimation to find an approximate value of $\dfrac{804}{39,500}$. Round each number so that there is one nonzero digit. Do not find the exact value.

Concept Check
Explain how you would change 0.0078 to a percent.

Name: _____   Date: _____

Instructor: _____   Section: _____

Chapter 0 Prealgebra Review
0.6 Using the Mathematics Blueprint for Problem Solving

Vocabulary
mathematics blueprint   •   area   •   percent   •   check   •   estimation

1. When solving real-life problems, use a(n) _____.

2. You can _____ your answer to real-life problems using estimation.

Example	Student Practice
1. Nancy and John want to install wall-to-wall carpeting in their living room. The floor of the rectangular living room is $11\frac{2}{3}$ feet wide and $19\frac{1}{2}$ feet long. How much will it cost if the carpet is $18.00 per square yard?	**2.** Kris wants to install wall-to-wall carpet in a bedroom. The floor of this rectangular room is $12\frac{1}{2}$ feet wide and $17\frac{1}{3}$ feet long. How much will it cost if the carpet is $27.00 per square yard?

Read the problem carefully and create a Mathematics Blueprint. Draw a picture if it will help.

Find the area of the floor.

$$11\frac{2}{3}\times19\frac{1}{2}=\frac{35}{3}\times\frac{39}{2}=\frac{455}{2}=227\frac{1}{2}$$

A minimum of $227\frac{1}{2}$ square feet of carpet is needed. Dividing this value by 9 gives the area in square yards,

$$227\frac{1}{2}\div 9=25\frac{5}{18}.$$ Multiply $25\frac{5}{18}$ square yards by $18.00 per square yard to find the cost.

$$25\frac{5}{18}\times18=\frac{455}{18}\times\frac{18}{1}=455$$

The carpet will cost a minimum of $455.00 for this room. The check is left to the student.

Vocabulary Answers: 1. mathematics blueprint 2. check

Example	Student Practice
3. The following chart shows the 2012 sales of Micropower Computer Software for each of the four regions of the United States. Use the chart to answer the following questions (round all answers to the nearest whole percent).	**4.** Use the chart in example **3** to answer the following.

Region of the U.S.	Number of Sales Personnel	Dollar Volume of Sales
Northeast	12	1,560,000
Southeast	18	4,300,000
Northwest	10	3,660,000
Southwest	15	3,720,000
Total	55	13,240,000

(a) What percent of sales personnel are assigned to the Northwest?

(a) What percent of the sales personnel are assigned to the Northeast?

Read the problem carefully and create a Mathematics Blueprint. You must calculate, "12 is what percent of 55?" Divide 12 by 55. Change the decimal to a percent and round.

$$\frac{12}{55} = 0.21818... = 21.818...\% \approx 22\%$$

Thus, about 22% of sales personnel are assigned to the Northeast.

(b) What percent of the volume of sales is attributed to the Northwest?

(b) What percent of the volume of sales is attributed to the Northeast?

You must calculate, "1,560,000 is what percent of 13,240,000?"

$$\frac{1,560,000}{13,240,000} = \frac{156}{1324} \approx 0.1178 \approx 12\%$$

Thus, about 12% of the volume of sales is attributed to the Northeast.

(c) Of the Northeast and Northwest, which region has sales personnel that appear to be more effective in terms of the volume of sales?

Extra Practice

1. In a recent mayoral election in the town of Oceanside, Ms. Dempsey received 44% of the vote, Mr. Chan received 36%, and Ms. Tobin received 18%. A total of 6150 people voted. How many more people voted for Ms. Dempsey than Mr. Chan?

2. Ally would like to install linoleum on her kitchen floor. The floor of the rectangular kitchen is $16\frac{1}{2}$ feet wide and $22\frac{1}{2}$ feet long. How much will it cost to replace the floor if the linoleum costs $17.00 per square yard?

3. Nadine ran 2.2 kilometers on Monday. On each of the next four days, she increased her distance by 40% compared with the previous day. How many kilometers further did Nadine run on Thursday than on Wednesday? Round to the nearest tenth of a kilometer.

4. Karen earns $5200 per month, of which 35% is taken out in taxes and other withholding. How much does Karen take home every month after taxes and withholding?

Concept Check

Hank knows that 1 kilometer ≈ 0.62 mile. Explain how Hank could find out how many miles he traveled on a 12-kilometer trip to Mexico.

MATH COACH

Mastering the skills you need to do well on the test.

Watch the **MATH COACH** videos in MyMathLab® or on YouTube™ while you work the problems below. These helpful hints will help you avoid making common errors on test problems.

Subtracting Mixed Numbers—Problem 7

Subtract $3\frac{2}{3} - 2\frac{5}{6}$.

> **Helpful Hint:** First change the mixed numbers to improper fractions. Next find the LCD of the two denominators. Then change the fractions to an equivalent form with the LCD as the common denominator before subtracting.

Did you change $3\frac{2}{3}$ to $\frac{11}{3}$ and $2\frac{5}{6}$ to $\frac{17}{6}$?

Yes _____ No _____

If you answered No, stop and change the two mixed numbers to improper fractions.

Did you find the LCD to be 6? Yes _____ No _____

If you answered No, consider how to find the LCD of the two denominators. Once the fractions are written as equivalent fractions with 6 as the denominator, the two like fractions can be subtracted.

If you answered Problem 7 incorrectly, go back and rework the problem using these suggestions.

Dividing Mixed Numbers—Problem 10 Divide $5\frac{3}{8} \div 2\frac{3}{4}$.

> **Helpful Hint:** Be sure to change the mixed numbers to improper fractions before dividing.

Did you change $5\frac{3}{8}$ to $\frac{43}{8}$ and $2\frac{3}{4}$ to $\frac{11}{4}$ before doing any other steps? Yes _____ No _____

If you answered No, stop and change the two mixed numbers to improper fractions.

Next did you change the division to multiplication to obtain $\frac{43}{8} \times \frac{4}{11}$? Yes _____ No _____

If you answered No, stop and make this change.

Did you simplify the product? Yes _____ No _____

If you answered No, try dividing a 4 from the second numerator and first denominator before multiplying. The product will be an improper fraction that can be converted to a mixed number.

Now go back and rework the problem using these suggestions.

Dividing Decimals—Problem 17   Divide $12.88 \div 0.056$.

> **Helpful Hint:** Be careful as you move the decimal point in the divisor to the right. Make sure that the resulting divisor is an integer. Then move the decimal point in the dividend the same number of places to the right. Add zeros if necessary.

Did you move the decimal point in the divisor three places to the right to get 56?

Yes _____   No _____

If you answered No, stop and perform this step first.

Did you move the decimal point in the dividend three places to the right and add one zero to get 12880?

Yes _____   No _____

If you answered No, perform this step now. Be careful of calculation errors as you perform the division.

If you answered problem 17 incorrectly, go back and rework the problem using these suggestions.

Find the Missing Percent—Problem 22   39 is what percent of 650?

> **Helpful Hint:** Write a fraction with the two numbers. The number after the word "of" is always the denominator, and the other number is the numerator.

Did you write the fraction $\dfrac{39}{650}$?

Yes _____   No _____

If you answered No, stop and perform this step.

Did you simplify the fraction to $\dfrac{3}{50}$ before changing the fraction to a decimal?

Yes _____   No _____

If you answered No, consider that simplifying the fraction first makes the division step a little easier. Be sure to place the decimal point correctly in your quotient.

Did you change the quotient from a decimal to a percent?

Yes _____   No _____

If you answered No, stop and perform this final step.

Now go back and rework the problem using these suggestions.

Chapter 1 Real Numbers and Variables
1.1 Adding Real Numbers

Vocabulary
whole numbers • integers • rational numbers • irrational numbers • real numbers
number line • positive numbers • negative numbers • opposite numbers • absolute value
additive inverses

1. The _____ of a number is the distance between that number and zero on a
    number line.

2. _____, also called additive inverses, have the same magnitude but different signs
    and can be represented on a number line.

3. _____ are all the rational numbers and all the irrational numbers.

4. _____ are to the left of 0 on a number line.

Example	**Student Practice**
1. Classify as an integer, a rational number, an irrational number, and/or a real number.	**2.** Classify as an integer, a rational number, an irrational number, and/or a real number.
(a) 5	**(a)** $\sqrt{10}$
5 is an integer, a rational number, and a real number.	
(b) $-\dfrac{1}{3}$	**(b)** $\dfrac{1}{9}$
$-\dfrac{1}{3}$ is a rational number and a real number.	
(c) $\sqrt{2}$ is an irrational number and a real number.	**(c)** -4.25

Vocabulary Answers: 1. absolute value 2. opposite numbers 3. real numbers 4. negative numbers

Example	Student Practice
3. Use a real number to represent each situation.	**4.** Use a real number to represent each situation.
(a) A temperature of 128.6° F below zero is recorded at Vostok, Antarctica.	(a) A stock loss of 6.23 points
"below" is a keyword indicating that the number is negative.	
128.6° F below zero is −128.6.	(b) A temperature of 109.4° F in a desert
(b) The Himalayan peak K2 rises 29,064 feet above sea level.	
"above" is a keyword indicating that the number is positive.	(c) A population loss of 345
29,064 feet above sea level is +29,064.	
(c) The Dow gains 10.24 points.	(d) A 15 yard gain in a football game
A gain of 10.24 points is +10.24.	
(d) An oil drilling platform extends 328 feet below sea level.	
328 feet below sea level is −328.	
5. Find the additive inverse (that is, the opposite).	**6.** Find the additive inverse (that is, the opposite).
(a) −7	(a) $-\dfrac{7}{8}$
The opposite of −7 is +7.	
(b) $\dfrac{1}{4}$	(b) 30 feet below sea level
The opposite of $\dfrac{1}{4}$ is $-\dfrac{1}{4}$.	

28

Example	Student Practice
7. Find the absolute value.	**8.** Find the absolute value.
(a) $\|-4.62\|$	**(a)** $\|5.34\|$
$\|-4.62\| = 4.62$	
(b) $\left\|\dfrac{3}{7}\right\|$	**(b)** $\left\|-\sqrt{5}\right\|$
$\left\|\dfrac{3}{7}\right\| = \dfrac{3}{7}$	
(c) $\|0\|$	**(c)** $\left\|\dfrac{0}{6}\right\|$
$\|0\| = 0$	

9. Add. $14+6$	**10.** Add. $-9+(-5)$
To add two numbers with the same sign, add the absolute values. Then use the common sign, $+$, in the answer.	
$14+16 = 30$	
$14+16 = +30$	

11. Add. $8+(-7)$	**12.** Add. $3+(-9)$
To add two numbers with different signs, find the difference between the two absolute values. The answer will have the sign of the number with the larger absolute value.	
$8-7 = 1$	
$+8+(-7) = +1 \text{ or } 1$	

Example	Student Practice
13. Add. $-1.8 + 1.4 + (-2.6)$	**14.** Add. $5.6 + (-2.34) + (-3.16)$
We take the difference of 1.8 and 1.4 and use the sign of the number with the larger absolute value. Then add the result to -2.6. $$-0.4 + (-2.6) = -3.0$$	
15. Add. $-8 + 3 + (-5) + (-2) + 6 + 5$	**16.** Add. $-7 + 2 + 5 + (-6) + 1 + (-3)$

15. Add the three negative numbers and the three positive numbers separately, then add the two results.

$$
\begin{array}{rr}
-8 & +3 \\
-5 & +6 \\
\underline{-2} & \underline{+5} \\
-15 & +14
\end{array}
$$

Add the two results, $-15 + 14 = -1$.

Extra Practice

1. Identify the following as a whole number, rational number, irrational number, integer, or real number. Remember, a number can be identified as more than one type.

π

2. Find the absolute value of $-\dfrac{7}{8}$.

3. Use a real number to represent the situation.

You owe your best friend $25.

4. Add. $57 + (-32) + 90 + (-100)$

Concept Check

Explain why when you add two negative numbers, you always obtain a negative number, but when you add one negative number and one positive number, you may obtain zero, a positive number, or a negative number.

Chapter 1 Real Numbers and Variables
1.2 Subtracting Real Numbers

Vocabulary
additive inverse property   •   subtract

1. To _____ real numbers, add the opposite of the second number to the first.

2. The _____ says that when you add two real numbers that are opposites of each other, you will obtain zero.

Example	Student Practice
1. Subtract. $6-(-2)$	**2.** Subtract. $4-(-8)$
Change the subtraction to addition and write the opposite of the second number.	
$6-(-2)=6+(+2)$	
Then add the two real numbers with the same sign.	
$6+(+2)=8$	
3. Subtract. $-8-(-6)$	**4.** Subtract. $-15-(-9)$
Change the subtraction to addition and write the opposite of the second number.	
$-8-(-6)=-8+(+6)$	
Then add the two real numbers with the same sign.	
$-8+(+6)=-2$	

Vocabulary Answers: 1. subtract 2. additive inverse property

Example	Student Practice
5. Subtract.	**6.** Subtract.

5. Subtract.

(a) $\dfrac{3}{7} - \dfrac{6}{7}$

Note that the problem has two fractions with the same denominator.

$$\frac{3}{7} - \frac{6}{7} = \frac{3}{7} + \left(-\frac{6}{7}\right)$$

$$= -\frac{3}{7}$$

(b) $-\dfrac{7}{18} - \left(-\dfrac{1}{9}\right)$

Change subtracting to adding the opposite and change $\dfrac{1}{9}$ to $\dfrac{2}{18}$ since the LCD $= 18$.

$$-\frac{7}{18} - \left(-\frac{1}{9}\right) = -\frac{7}{18} + \frac{1}{9}$$

$$= -\frac{7}{18} + \frac{2}{18}$$

$$= -\frac{5}{18}$$

6. Subtract.

(a) $\dfrac{2}{9} - \dfrac{7}{9}$

(b) $-\dfrac{2}{5} - \left(-\dfrac{1}{4}\right)$

---

7. Subtract. $-5.2 - (-5.2)$

Change the subtraction problem to one of adding the opposite of the second number. Then, add two numbers with different signs.

$$-5.2 - (-5.2) = -5.2 + 5.2$$
$$= 0$$

8. Subtract. $-4.2 - (-4.2)$

Example	Student Practice
9. Subtract.	**10.** Subtract.

(a) $-8-2$

To subtract, we add the opposite of the second number to the first.

$$-8-2=-8+(-2)=-10$$

(b) $23-28$

In a similar fashion, we have the following.

$$23-28=23+(-28)=-5$$

(c) $5-(-3)$

$$5-(-3)=5+3=8$$

(d) $\dfrac{1}{4}-8$

Write -8 as $-\dfrac{32}{4}$ since the LCD $=4$.

$$\dfrac{1}{4}-8=\dfrac{1}{4}+(-8)$$
$$=\dfrac{1}{4}+\left(-\dfrac{32}{4}\right)$$
$$=-\dfrac{31}{4}\ \text{or}\ -7\dfrac{3}{4}$$

(a) $-12-8$

(b) $15-20$

(c) $14-(-11)$

(d) $\dfrac{4}{5}-6$

Example	Student Practice
11. A satellite is recording radioactive emissions from nuclear waste buried 3 miles below sea level. The satellite orbits Earth at 98 miles above sea level. How far is the satellite from the nuclear waste? We want to find the difference between +98 miles and −3 miles. This means we must subtract −3 from 98. $$98 - (-3) = 93 + 3 = 101$$ The satellite is 101 miles from the nuclear waste.	**12.** A helicopter flies over a deep sea diver. The helicopter is 500 feet above sea level. The diver is 129 feet below sea level. How far is the helicopter from the diver?

Extra Practice

1. Subtract by adding the opposite.

$$(-80.09) - (-76.8)$$

2. Subtract by adding the opposite.

$$-\frac{11}{15} - \left(-\frac{1}{2}\right)$$

3. Change each subtraction operation to "adding the opposite." Then combine the numbers.

$$-30 + 12 - (-15) - 8$$

4. In the morning, Maxine began hiking uphill from a valley that is 22 feet below sea level. She ate her lunch on top of a hill whose altitude is 179 feet above sea level. Write an expression to represent the difference in altitude. How many feet did Maxine climb during her hike?

Concept Check

Explain the different results that are possible when you start with a negative number and then subtract a negative number.

Chapter 1 Real Numbers and Variables
1.3 Multiplying and Dividing Real Numbers

Vocabulary

multiplication • division • positive • negative • undefined

1. Division by zero is _____.

2. When you multiply or divide two numbers with different signs, you obtain a _____ number.

3. _____ can be indicated by the symbol ÷ or by the fraction bar − .

4. _____ is commutative and associative.

Example	Student Practice
1. Multiply.	**2.** Multiply.
(a) $\left(-\dfrac{5}{7}\right)\left(-\dfrac{2}{9}\right)$	**(a)** $(5)(4)$
When multiplying two numbers with the same sign, the result is a positive number.	
$\left(-\dfrac{5}{7}\right)\left(-\dfrac{2}{9}\right) = \dfrac{10}{63}$	
(b) $-4(8)$	**(b)** $\left(-\dfrac{5}{12}\right)(5)$
When multiplying two numbers with different signs, the result is a negative number.	
$-4(8) = -32$	

Vocabulary Answers: 1. undefined 2. negative 3. division 4. multiplication

Example	Student Practice
3. Multiply.	**4.** Multiply.
(a) $\left(-\dfrac{1}{2}\right)(-1)(-4)$	**(a)** $-6(-5.4)$
Begin by multiplying the first two numbers. The signs are the same, so the answer is positive. Then multiply the remaining two numbers. The signs are different, so the answer is negative.	
$\left(-\dfrac{1}{2}\right)(-1)(-4) = +\dfrac{1}{2}(-4) = -2$	**(b)** $-3(-2)(-5)(-1)(-2)$
(b) $-2(-2)(-2)(-2)$	
$\begin{aligned} -2(-2)(-2)(-2) &= +4(-2)(-2) \\ &= -8(-2) \\ &= +16 \text{ or } 16 \end{aligned}$	
5. Divide.	**6.** Divide.
(a) $12 \div 4$	**(a)** $(-32) \div (-4)$
When dividing two numbers with the same sign, the result is a positive number.	
$12 \div 4 = 3$	
(b) $\dfrac{-36}{18}$	**(b)** $\dfrac{45}{-9}$
When dividing two numbers with different signs, the result is a negative number.	
$-\dfrac{36}{18} = -2$	

Example	Student Practice

7. Divide.

(a) $-36 \div 0.12$

When dividing two numbers with different signs, the result is a negative number. We then divide the absolute values.

$$0.12 \wedge \overline{)36.00 \wedge}$$

with quotient

$$\begin{array}{r} 3\ 00. \\ 0.12_\wedge\overline{)36.00_\wedge} \\ \underline{36} \\ 00 \end{array}$$

Thus $-36 \div 0.12 = -300$.

(b) $-2.4 \div (-0.6)$

$$\begin{array}{r} 4. \\ 0.6_\wedge\overline{)2.4_\wedge} \\ \underline{2\ 4} \end{array}$$

Thus $-2.4 \div (-0.6) = 4$.

8. Divide.

(a) $-121 \div (-0.11)$

(b) $-0.54 \div 0.9$

---

9. Divide. $-\dfrac{12}{5} \div \dfrac{2}{3}$

We invert the second fraction and multiply by the first fraction. The answer is negative since the two numbers divided have different signs.

$$-\frac{12}{5} \div \frac{2}{3} = \left(-\frac{12}{5}\right)\left(\frac{3}{2}\right)$$

$$= \left(-\frac{\overset{6}{\cancel{12}}}{5}\right)\left(\frac{3}{\underset{1}{\cancel{2}}}\right)$$

$$= -\frac{18}{5} \text{ or } -3\frac{3}{5}$$

10. Divide. $-\dfrac{2}{9} \div \left(-\dfrac{8}{15}\right)$

Example	Student Practice
11. Divide. $\dfrac{-\dfrac{2}{3}}{-\dfrac{7}{13}}$	**12.** Divide. $\dfrac{\dfrac{32}{4}}{-\dfrac{4}{5}}$

$$\dfrac{-\dfrac{2}{3}}{-\dfrac{7}{13}} = -\dfrac{2}{3} \div \left(-\dfrac{7}{13}\right)$$

$$= -\dfrac{2}{3}\left(-\dfrac{13}{7}\right)$$

$$= \dfrac{26}{21} \text{ or } 1\dfrac{5}{21}$$

Extra Practice

1. Multiply. Be sure to write your answer in the simplest form.

$$(15.8)(-29.3)$$

2. Multiply. You may want to determine the sign of the product before you multiply.

$$(2.5)(-0.4)(-3.2)(5)$$

3. Divide.

$$87.5 \div (-0.5)$$

4. Divide.

$$-\dfrac{11}{16} \div \dfrac{33}{40}$$

Concept Check

Explain how you can determine the sign of the answer if you multiply several negative numbers.

Chapter 1 Real Numbers and Variables
1.4 Exponents

Vocabulary

base • exponent • variable • squared • cubed • to the (exponent)-th power

1. The _____ tells you how many times the base is used as a factor.

2. If the base has an exponent of 3, we say the base is _____.

3. The _____ tells you what number is being multiplied in exponent form.

4. If we do not know the value of a number, we use a letter, called a(n) _____, to represent the unknown number.

Example	**Student Practice**
1. Write in exponent form.	**2.** Write in exponent form.
(a) $9(9)(9)$	**(a)** $-3(-3)(-3)(-3)$
The base is 9 and it is used as a factor three times. So, the exponent is 3.	
$9(9)(9) = 9^3$	
(b) $-7(-7)(-7)(-7)(-7)$	**(b)** $-6(-6)(-6)(-6)(-6)(-6)(-6)(-6)$
The base is -7 and it is used as a factor five times. The answer must contain parentheses.	
$-7(-7)(-7)(-7)(-7) = (-7)^5$	**(c)** $(n)(n)(n)(n)(n)(n)$
(c) $(y)(y)(y)$	
$(y)(y)(y) = y^3$	

Vocabulary Answers: 1. exponent 2. cubed 3. base 4. variable

Example	Student Practice
3. Evaluate.	**4.** Evaluate.
(a) 2^5	**(a)** 4^3
$2^5 = (2)(2)(2)(2)(2) = 32$	
(b) $2^3 + 4^4$	
First evaluate each power. Then add.	**(b)** $1^7 + 6^2$
$\begin{aligned} 2^3 + 4^4 &= 8 + 256 \\ &= 264 \end{aligned}$	
5. Evaluate.	**6.** Evaluate.
(a) $(-2)^3$	**(a)** $(-4)^3$
The answer is negative since the base is negative and the exponent is odd, $(-2)^3 = -8$.	
(b) $(-4)^6$	**(b)** $(-4)^4$
The answer is positive since the exponent 6 is even, $(-4)^6 = +4096$.	
(c) -3^6	**(c)** -4^4
The negative sign is not contained within parentheses. Thus, find 3^6 and take the opposite of that value, $-3^6 = -729$.	
(d) $-(5^4)$	**(d)** $-(4^4)$
The negative sign is outside the parentheses, $-(5^4) = -625$.	

Example	Student Practice
7. Evaluate.	**8.** Evaluate.

7. Evaluate.

(a) $\left(\dfrac{1}{2}\right)^4$

$$\left(\dfrac{1}{2}\right)^4 = \left(\dfrac{1}{2}\right)\left(\dfrac{1}{2}\right)\left(\dfrac{1}{2}\right)\left(\dfrac{1}{2}\right)$$
$$= \dfrac{1}{16}$$

(b) $(0.2)^4$

$$(0.2)^4 = (0.2)(0.2)(0.2)(0.2)$$
$$= 0.0016$$

(c) $\left(\dfrac{2}{5}\right)^3$

$$\left(\dfrac{2}{5}\right)^3 = \left(\dfrac{2}{5}\right)\left(\dfrac{2}{5}\right)\left(\dfrac{2}{5}\right)$$
$$= \dfrac{8}{125}$$

(d) $(3)^3(2)^5$

First we evaluate each power, then we multiply.

$$(3)^3(2)^5 = (27)(32)$$
$$= 864$$

(e) $2^3 - 3^4$

$$2^3 - 3^4 = 8 - 81$$
$$= -73$$

8. Evaluate.

(a) $\left(\dfrac{1}{5}\right)^3$

(b) $(0.3)^3$

(c) $\left(\dfrac{4}{7}\right)^3$

(d) $(2)^4(5)^2$

(e) $10^3 - 2^{10}$

Extra Practice

1. Write the product in exponent form. Do not evaluate.

$$(-ab)(-ab)$$

2. Evaluate.

$$\left(-\frac{1}{2}\right)^3$$

3. Evaluate.

$$-6^2 - (-2)^2$$

4. Evaluate.

$$7^2 - (-2)^3$$

Concept Check

Explain the difference between $(-2)^6$ and -2^6. How do you decide if the answers are positive or negative?

Chapter 1 Real Numbers and Variables
1.5 The Order of Operations

Vocabulary
order of operations   •   parentheses   •   multiply and divide   •   add and subtract

1. When simplifying an expression, the last priority is to _____ numbers from left to right.

2. When simplifying an expression, the first priority is to do all operations inside _____.

3. The list of operations to do first is called the _____.

4. When simplifying an expression, the third priority is to _____ numbers from left to right.

Example	Student Practice
1. Evaluate. $8 \div 2 \cdot 3 + 4^2$	**2.** Evaluate. $36 \div 3 \cdot 2 + 6^2$

1. Evaluate. $8 \div 2 \cdot 3 + 4^2$

Evaluate $4^2 = 16$ because the highest priority in this problem is raising to a power.

$8 \div 2 \cdot 3 + 4^2 = 8 \div 2 \cdot 3 + 16$

Next, multiply and divide from left to right.

$8 \div 2 \cdot 3 + 16 = 4 \cdot 3 + 16$
$\qquad\qquad\quad = 12 + 16$

Finally, add.

$12 + 16 = 28$

Vocabulary Answers: 1. add and subtract 2. parentheses 3. order of operations 4. multiply and divide

Example	Student Practice
3. Evaluate. $(-3)^3 - 2^4$	**4.** Evaluate. $(-5)^2 - 1^{10}$

3. Evaluate. $(-3)^3 - 2^4$

The highest priority is to raise the expressions to the appropriate powers.

In $(-3)^3$ we are cubing the number -3 to obtain -27.

$$(-3)^3 - 2^4 = -27 - 2^4$$

Be careful; -2^4 is not $(-2)^4$. We raise 2 to the fourth power. Then, add and subtract from left to right.

$$(-3)^3 - 2^4 = -27 - 16$$
$$= -43$$

4. Evaluate. $(-5)^2 - 1^{10}$

---

5. Evaluate. $2 \cdot (2-3)^3 + 6 \div 3 + (8-5)^2$

Combine the numbers inside the parentheses.

$$2 \cdot (2-3)^3 + 6 \div 3 + (8-5)^2$$
$$= 2 \cdot (-1)^3 + 6 \div 3 + 3^2$$

We need parentheses for -1 because of the negative sign, but they are not needed for 3. Next, raise to a power.

$$2 \cdot (-1)^3 + 6 \div 3 + 3^2 = 2 \cdot (-1) + 6 \div 3 + 9$$

Next, multiply and divide from left to right, $2 \cdot (-1) + 6 \div 3 + 9 = -2 + 2 + 9$.

Finally, add and subtract from left to right, $-2 + 2 + 9 = 9$.

6. Evaluate. $3 \cdot (3-5)^2 + 15 \div 5 + (7-5)^3$

Example	Student Practice

7. Evaluate. $\left(-\dfrac{1}{5}\right)\left(\dfrac{1}{2}\right)-\left(\dfrac{3}{2}\right)^2$

The highest priority is to raise $\dfrac{3}{2}$ to the second power, $\left(\dfrac{3}{2}\right)^2 = \left(\dfrac{3}{2}\right)\left(\dfrac{3}{2}\right)=\dfrac{9}{4}$.

$$\left(-\dfrac{1}{5}\right)\left(\dfrac{1}{2}\right)-\left(\dfrac{3}{2}\right)^2 = \left(-\dfrac{1}{5}\right)\left(\dfrac{1}{2}\right)-\dfrac{9}{4}$$

Next we multiply.

$$\left(-\dfrac{1}{5}\right)\left(\dfrac{1}{2}\right)-\dfrac{9}{4} = -\dfrac{1}{10}-\dfrac{9}{4}$$

We need to write each fraction as an equivalent fraction with the LCD of 20.

$$\left(-\dfrac{1}{5}\right)\left(\dfrac{1}{2}\right)-\dfrac{9}{4} = -\dfrac{1\cdot 2}{10\cdot 2}-\dfrac{9\cdot 5}{4\cdot 5}$$

$$= -\dfrac{2}{20}-\dfrac{45}{20}$$

Finally, subtract.

$$-\dfrac{2}{20}-\dfrac{45}{20}=-\dfrac{47}{20} \text{ or } -2\dfrac{7}{20}$$

8. Evaluate. $\left(\dfrac{3}{4}\right)^2 + \dfrac{14}{15}\left(-\dfrac{5}{7}\right)$

Extra Practice

1. Evaluate. Simplify all fractions.

   $1 + 2 - 3 \times 4 \div 6$

2. Evaluate. Simplify all fractions.

   $50 \div 5 \times 2 + (11 - 8)^2$

3. Evaluate. Simplify all fractions.

   $\left(\dfrac{1}{2}\right)^2 \div \left(\dfrac{1}{2}\right)^3$

4. Evaluate. Simplify all fractions.

   $\left(\dfrac{2}{3}\right)^2 (-18) + \dfrac{3}{7} \div \dfrac{5}{21}$

Concept Check

Explain in what order you would perform the calculations to evaluate the expression $4 - (-3 + 4)^3 + 12 \div (-3)$.

Chapter 1 Real Numbers and Variables
1.6 Using the Distributive Property to Simplify Algebraic Expressions

Vocabulary
algebraic expression • term • distributive property • factors

1. The _____ states that for all real numbers a, b, and c, $a(b+c)=ab+ac$.

2. A(n) _____ is a quantity that contains numbers and variables.

3. Two or more algebraic expressions that are multiplied are called _____.

4. A(n) _____ is a number, variable, or a product of numbers and variables.

Example	Student Practice
1. Multiply.	**2.** Multiply.
(a) $5(a+b)$	**(a)** $6(x+3y)$
Multiply the factor $(a+b)$ by the factor 5 using the distributive property.	
$5(a+b)=5a+5b$	**(b)** $-4(2m+5n)$
(b) $-3(3x+2y)$	
$-3(3x+2y)=-3(3x)+(-3)(2y)$ $\qquad = -9x-6y$	
3. Multiply. $-(a-2b)$	**4.** Multiply. $-(m-6n)$
$-(a-2b)=(-1)(a-2b)$ $\qquad = (-1)(a)+(-1)(-2b)$ $\qquad = -a+2b$	

Vocabulary Answers: 1. distributive property 2. algebraic expression 3. factor 4. term

Example	Student Practice
5. Multiply.	**6.** Multiply.
(a) $\frac{2}{3}\left(x^2 - 6x + 8\right)$	**(a)** $\frac{3}{4}\left(m^2 - 8m + 5\right)$
Apply the distributive property.	
$\frac{2}{3}\left(x^2 - 6x + 8\right)$	
$= \left(\frac{2}{3}\right)\left(1x^2\right) + \left(\frac{2}{3}\right)\left(-6x\right) + \left(\frac{2}{3}\right)(8)$	
$= \frac{2}{3}x^2 + \left(-4x\right) + \frac{16}{3}$	**(b)** $1.3\left(x^2 + 1.3x + 0.7\right)$
$= \frac{2}{3}x^2 - 4x + \frac{16}{3}$	
(b) $1.4\left(a^2 + 2.5a + 1.8\right)$	
$1.4\left(a^2 + 2.5a + 1.8\right)$	
$= 1.4\left(1a^2\right) + (1.4)(2.5a) + (1.4)(1.8)$	
$= 1.4a^2 + 3.5a + 2.52$	
7. Multiply. $-2x\left(3x + y - 4\right)$	**8.** Multiply. $-5x\left(x + 3y - 6\right)$
$-2x\left(3x + y - 4\right)$	
$= -2(x)(3)(x) - 2(x)(y) - 2(x)(-4)$	
$= -2(3)(x)(x) - 2(xy) - 2(-4)(x)$	
$= -6x^2 - 2xy + 8x$	
9. Multiply. $\left(2x^2 - x\right)(-3)$	**10.** Multiply. $\left(4y - 3y^2\right)(-2)$
$\left(2x^2 - x\right)(-3) = 2x^2(-3) + (-x)(-3)$	
$\qquad\qquad = -6x^2 + 3x$	

Example	Student Practice

11. A farmer has a rectangular field that is 300 feet wide. One portion of the field is $2x$ feet long. The other portion of the field is $3y$ long. Use the distributive property to find an expression for the area of this field.

First we draw a picture of a field that is 300 feet wide and $2x + 3y$ feet long.

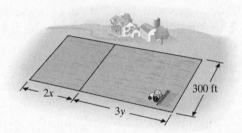

To find the area of the field, we multiply the width times the length.

$$300(2x + 3y)$$

Then, apply the distributive property.

$$300(2x + 3y) = 300(2x) + 300(3y)$$

Finally, simplify.

$$300(2x) + 300(3y) = 600x + 900y$$

Thus the area of the field in square feet is $600x + 900y$.

12. A convenience store is 500 feet wide. One portion of the store is $7x$ feet long. The other portion of the store is $8y$ feet long. Use the distributive property to find an expression for the area of this store.

Extra Practice

1. Multiply. Use the distributive property.

$$-2(6x+3)$$

2. Multiply. Use the distributive property.

$$\frac{1}{2}\left(\frac{1}{3}x+\frac{2}{5}y\right)$$

3. Multiply. Use the distributive property.

$$(6x-4y+2z)(-3)$$

4. Mr. Jorgensen's backyard is a rectangle that is 30 feet wide. He recently built a fence that extends across his backyard. The distance from his house to the fence is 15 feet. The distance from the fence to the back edge of his property is $7n$. Write an expression that represents the area of Mr. Jorgensen's property in square feet, then simplify this expression.

Concept Check

Multiply. Use the distributive property and explain how you would multiply to obtain the answer for $\left(-\frac{3}{7}\right)(21x^2-14x+3)$.

Name: _____   Date: _____

Instructor: _____   Section: _____

Chapter 1 Real Numbers and Variables
1.7 Combining Like Terms

Vocabulary

combine like terms   •   like terms   •   combining   •   term

1. A _____ is a number, a variable, or a product of numbers and variables separated by plus or minus signs in an algebraic expression.

2. We can add or subtract quantities that are like quantities, which is called _____ like quantities.

3. To _____ is to add or subtract like terms in an algebraic expression.

4. _____ are terms that have identical variables and exponents.

Example	Student Practice
1. List the like terms of each expression.	**2.** List the like terms of each expression.
(a) $5x - 2y + 6x$	**(a)** $4x - 6y + 2x + 7$
$5x$ and $6x$ are like terms.	
(b) $2x^2 - 3x - 5x^2 - 8x$	**(b)** $4x^2 + 5x - 6x^2 + 9x$
$2x^2$ and $-5x^2$ are like terms. $-3x$ and $-8x$ are like terms. Note that x^2 and x are not like terms.	
3. Combine like terms.	**4.** Combine like terms.
(a) $-4x^2 + 8x^2$	**(a)** $10z^5 - 3z^5$
Each term contains the factor x^2.	
$-4x^2 + 8x^2 = (-4 + 8)x^2 = 4x^2$	
	(b) $5z - 7z + 4z$
(b) $5x + 3x + 2x$	
$5x + 3x + 2x = (5 + 3 + 2)x = 10x$	

Vocabulary Answers: 1. term 2. combining 3. combine like terms 4. like terms

Example	Student Practice
5. Simplify. $5a^2 - 2a^2 + 6a^2$	**6.** Simplify. $x^3 - 7x^3 + 9x^3$

$$5a^2 - 2a^2 + 6a^2 = (5 - 2 + 6)a^2$$
$$= 9a^2$$

---

7. Simplify.

  (a) $5.6a + 2b + 7.3a - 6b$

  We combine the a terms and the b terms separately.

  $5.6a + 2b + 7.3a - 6b = 12.9a - 4b$

  (b) $3x^2y - 2xy^2 + 6x^2y$

  Note that x^2y and xy^2 are not like terms because of different powers.

  $3x^2y - 2xy^2 + 6x^2y = 9x^2y - 2xy^2$

  (c) $2a^2b + 3ab^2 - 6a^2b^2 - 8ab$

  These terms cannot be combined; there are no like terms in this expression.

  $2a^2b + 3ab^2 - 6a^2b^2 - 8ab$

8. Simplify.

  (a) $4.1m + 6n + 3.5m + 1.2n$

  (b) $6a^2b - 4ab^3 + 2a^2b$

  (c) $6xy + 2x^2y - 9xy^2 + 5x^2y^2$

---

9. Simplify. $3a - 2b + 5a^2 + 6a - 8b - 12a^2$

There are three pairs of like terms. You can rearrange the terms so that like terms are together.

$$\underbrace{3a + 6a}_{a \text{ terms}} \underbrace{- 2b - 8b}_{b \text{ terms}} \underbrace{+ 5a^2 - 12a^2}_{a^2 \text{ terms}}$$
$$= 9a - 10b - 7a^2$$

10. Simplify. $9x - 9c + 2t + 4c - 4t + 5x$

Example	Student Practice
11. Simplify. $\dfrac{3}{4}x^2 - 5y - \dfrac{1}{8}x^2 + \dfrac{1}{3}y$	**12.** Simplify. $\dfrac{2}{5}x - \dfrac{6}{7}y + 3x + \dfrac{1}{2}y$

We need the least common denominator for the x^2 terms, which is 8.

$$\frac{3}{4}x^2 - \frac{1}{8}x^2 = \frac{3 \cdot 2}{4 \cdot 2}x^2 - \frac{1}{8}x^2$$
$$= \frac{6}{8}x^2 - \frac{1}{8}x^2$$
$$= \frac{5}{8}x^2$$

The least common denominator for the y terms is 3.

$$-\frac{5}{1}y + \frac{1}{3}y = \frac{-5 \cdot 3}{1 \cdot 3}y + \frac{1}{3}y$$
$$= \frac{-15}{3}y + \frac{1}{3}y$$
$$= -\frac{14}{3}y$$

Thus, our solution is $\dfrac{5}{8}x^2 - \dfrac{14}{3}y$.

13. Simplify. $6(2x + 3xy) - 8x(3 - 4y)$	**14.** Simplify. $5(2c + 7cd) - 3c(9 - 4d)$

First remove the parentheses using the distributive property. Then combine like terms.

$$6(2x + 3xy) - 8x(3 - 4y)$$
$$= 12x + 18xy - 24 + 32xy$$
$$= -12x + 50xy$$

Extra Practice

1. Combine like terms.

$$9a + 10 - 2a^2 - 11a + 2 + 6a^2$$

2. Combine like terms.

$$\frac{3}{7}x^2 - \frac{1}{2}y + \frac{3}{14}x^2 - \frac{1}{8}y$$

3. Simplify. Use the distributive property to remove parentheses; then combine like terms.

$$-2(7a - 3b) + 4(-2a + 4b)$$

4. Simplify. Use the distributive property to remove parentheses; then combine like terms.

$$2a(b - 4c) - 3c(-7a + b - 10d)$$

Concept Check

Explain how you would remove parentheses and then combine like terms to obtain the answer for $1.2(3.5x - 2.2y) - 4.5(2.0x + 1.5y)$.

Name: _____   Date: _____
Instructor: _____   Section: _____

Chapter 1 Real Numbers and Variables
1.8 Using Substitution to Evaluate Algebraic Expressions and Formulas

Vocabulary
evaluate • substituted • perimeter • area • right angles • altitude • rectangle
parallelogram • square • trapezoid • triangle • circle • circumference

1. A(n) _____ is a rectangle with all four sides equal.

2. _____ is the distance around a circle.

3. _____ is a measure of the amount of surface in a region.

4. You will use the order of operations to _____ variable expressions.

Example	**Student Practice**
1. Evaluate $\frac{2}{3}x - 5$ for $x = -6$.	**2.** Evaluate $\frac{3}{4}x - 7$ for $x = -24$.
Substitute -6 for x. $\frac{2}{3}x - 5 = \frac{2}{3}(-6) - 5$ $\phantom{\frac{2}{3}x - 5} = -4 - 5$ $\phantom{\frac{2}{3}x - 5} = -9$	

3. Evaluate for $x = -3$.
 	
 (a) $2x^2$
      Here the value x is squared.
      $2x^2 = 2(-3)^2$
         $= 2(9)$
         $= 18$
 (b) $(2x)^2$
      Here the value $(2x)$ is squared.
      $(2x)^2 = [2(-3)]^2$
         $= (-6)^2$
         $= 36$
	4. Evaluate for $x = -5$.
	
	 (a) $3y^2$
	
	 (b) $(3y)^2$

Vocabulary Answers: 1. square 2. circumference 3. area 4. evaluate

Example	Student Practice
5. Evaluate $x^2 + 3x$ for $x = -4$. Replace each x by -4. $\begin{aligned} x^2 + 3x &= (-4)^2 + 3(-4) \\ &= 16 + 3(-4) \\ &= 16 - 12 \\ &= 4 \end{aligned}$	**6.** Evaluate $5x^2 + 4x$ for $x = -3$.
7. Evaluate $x^3 + 2xy - 3x + 1$ for $x = 2$ and $y = -\dfrac{1}{4}$. Replace x with 2 and y with $-\dfrac{1}{4}$. $x^3 + 2xy - 3x + 1$ $= (2)^3 + 2(2)\left(-\dfrac{1}{4}\right) - 3(2) + 1$ $= 8 + (-1) - 6 + 1$ $= 8 + (-1) + (-6) + 1$ $= 9 + (-7)$ $= 2$	**8.** Evaluate $7x^2 - 10xy - 50y^2$ for $x = -4$ and $y = \dfrac{1}{5}$.
9. Find the area of a triangle with a base of 16 centimeters (cm) and a height of 12 centimeters (12). Substitute $a = 12$ cm and $b = 16$ cm in $A = \dfrac{1}{2}ab$. $A = \dfrac{1}{2}(12 \text{ cm})(16 \text{ cm})$ $= (6)(16)(\text{cm})^2$ $= 96$ square centimeters The area of the triangle is 96 square centimeters or 96 cm^2.	**10.** Find the area of a triangle with a height of 9 yards and a base of 3 yards.

Example	Student Practice
11. Find the area of a circle if the radius is 2 inches.	**12.** Find the area of a circle if the radius is 4 yards.

Write the formula and substitute the given values for the letters.

$$A = \pi r^2 \approx (3.14)(2 \text{ inches})^2$$

Raise to a power. Then multiply.

$$(3.14)(2 \text{ inches})^2 = (3.14)(4)(\text{in.})^2$$
$$= 12.56 \text{ in.}^2$$

Thus the area is 12.56 square inches or 12.56 in.^{2}

| **13.** What is the Celsius temperature when the Fahrenheit temperature is $F = -22°$? | **14.** What is the Celsius temperature when the Fahrenheit temperature is $F = -40°$? Use the formula $C = \dfrac{5}{9}(F - 32)$. |

Use the formula. Substitute -22 for F in the formula.

$$C = \frac{5}{9}(F - 32)$$
$$= \frac{5}{9}\left[(-22) - 32\right]$$

Combine the numbers inside the brackets. Then simplify and multiply.

$$\frac{5}{9}\left[(-22) - 32\right] = \frac{5}{9}(-54)$$
$$= (5)(-6)$$
$$= -30$$

The temperature is $-30°$ Celsius or $-30°\,\text{C}$.

Example	Student Practice
15. You are driving on a highway in Mexico. It has a posted maximum speed of 100 kilometers per hour. You are driving at 61 miles per hour. Are you exceeding the speed limit?	**16.** You are driving on a Canadian highway. The posted maximum speed limit is 110 kilometers per hour. You are driving at 70 miles per hour. Are you exceeding the speed limit?

Use the formula. Replace r by 61. Then multiply the numbers.

$$k \approx 1.61r$$
$$= (1.61)(61)$$
$$= 98.21$$

You are driving approximately 98 kilometers per hour. You are not exceeding the speed limit.

Extra Practice

1. Evaluate $\dfrac{2}{7}y - 7$ for $y = 14$.

2. Evaluate $2x^2 + 3x - 8$ for $x = 3$.

3. Evaluate $\dfrac{2a^2 - 3b}{b^2}$ for $a = -1$ and $b = 2$.

4. A circular patch of grass has a radius of 6 meters. What is the area of the patch of grass, rounded to the nearest square meter?

Concept Check

Explain how you would find the area of a circle if you know its diameter is 12 meters.

Chapter 1 Real Numbers and Variables
1.9 Grouping Symbols

Vocabulary

grouping symbols　　•　　fraction bars　　•　　distributive property　　•　　like terms

1. _____ are also considered grouping symbols.

2. Use the _____ and the rules for real numbers to remove grouping symbols

3. Many expressions in algebra use _____ such as parentheses, brackets, and braces.

Example	Student Practice
1. Simplify. $3\left[6-2(x+y)\right]$	**2.** Simplify. $4\left[6x-5(x+3)\right]$

We want to remove the innermost parentheses first. Therefore, we first use the distributive property to simplify.

$$3\left[6-2(x+y)\right]=3\left[6-2x-2y\right]$$

Use the distributive property again.

$$3\left[6-2x-2y\right]=18-6x-6y$$

3. Simplify. $-2\left[3a-(b+2c)+(d-3e)\right]$	**4.** Simplify. $-4\left[5x-(2y-7)+(3x-9)\right]$

Remove both inner sets of parentheses.

$$-2\left[3a-(b+2c)+(d-3e)\right]$$
$$=-2\left[3a-b-2c+d-3e\right]$$

Now remove the brackets by multiplying each term by -2.

$$-2\left[3a-b-2c+d-3e\right]$$
$$=-6a+2b+4c-2d+6e$$

Vocabulary Answers: 1. fraction bars 2. distributive property 3. grouping symbols

Example	Student Practice
5. Simplify. $$2\left[3x-(y+w)\right]-3\left[2x+2(3y-2w)\right]$$ $$= 2\left[3x-y-w\right]-3\left[2x+6y-4w\right]$$ $$= 6x-2y-2w-6x-18y+12w$$ $$= -20y+10w \text{ or } 10w-20y$$	**6.** Simplify. $$5\left[x-3(4-2x)\right]+(-3)\left[7x-6(x+1)\right]$$

7. Simplify. $-3\left\{7x-2\left[x-(2x-1)\right]\right\}$

Remember to remove the innermost grouping symbols first.

$$-3\left\{7x-2\left[x-(2x-1)\right]\right\}$$
$$= -3\left\{7x-2\left[x-2x+1\right]\right\}$$
$$= -3\left\{7x-2\left[-x+1\right]\right\}$$
$$= -3\left\{7x+2x-2\right\}$$
$$= -3\left\{9x-2\right\}$$
$$= -27x+6$$

8. Simplify. $-4\left\{6x-\left[3x-(2-x)\right]\right\}$

Extra Practice

1. Simplify. Remove grouping symbols and combine like terms. $-2a-3(a+3b)$

2. Simplify. Remove grouping symbols and combine like terms. $$-2\left[3(x-y)-2(x+y)\right]$$

3. Simplify. Remove grouping symbols and combine like terms. $$2b-\left\{3a+4\left[2a-(-2b+3a)\right]\right\}$$

4. Simplify. Remove grouping symbols and combine like terms. $$-2\left\{5x^2+2\left[3x-(5-x)\right]\right\}$$

Concept Check

Explain how you would simplify the following expression to combine like terms whenever possible. $3\left\{2-3\left[4x-2(x+3)+5x\right]\right\}$

MATH COACH

Mastering the skills you need to do well on the test.

Watch the **MATH COACH** videos in MyMathLab® or on You Tube™ while you work the problems below. These helpful hints will help you avoid making common errors on test problems.

Evaluating Exponential Expressions When the Base Is a Fraction—Problem 9 Simplify $\left(\dfrac{2}{3}\right)^4$.

Helpful Hint: Always write out the repeated multiplication.
- If a number is raised to the fourth power, then we write out the multiplication with that number appearing as a factor a total of four times.
- When the base is a fraction, this means that we must multiply the numerators and then multiply the denominators.

Did you rewrite the problem as $\left(\dfrac{2}{3}\right)\left(\dfrac{2}{3}\right)\left(\dfrac{2}{3}\right)\left(\dfrac{2}{3}\right)$?

Yes _____ No _____

If you answered No to this question, please complete this step now.

Did you multiply $2\times2\times2\times2$ in the numerator and $3\times3\times3\times3$ in the denominator? Yes _____ No _____

If you answered No, make this correction and complete the calculations.

If you answered Problem 9 incorrectly, go back and rework the problem using these suggestions.

Using the Order of Operations with Numerical Expressions—Problem 11

Simplify $3(4-6)^3+12\div(-4)+2$.

Helpful Hint: First, do all operations inside parentheses. Second, raise numbers to a power. Then multiply and divide numbers from left to right. As the last step, add and subtract numbers from left to right.

Did you first combine $4-6$ to obtain -2?
Yes _____ No _____

If you answered No, perform the operation inside the parentheses first.

Next, did you calculate $(-2)^3=-8$ before multiplying by 3?
Yes _____ No _____

If you answered No, go back and evaluate the exponential expression.

Did you multiply and divide before adding?
Yes _____ No _____

If you answered No, remember that multiplication and division must be performed before addition and subtraction once the operations inside the parentheses are done and exponents are evaluated.

Evaluate Algebraic Expressions for a Specified Value—Problem 19

Evaluate $3x^2 - 7x - 11$ for $x = -3$.

Helpful Hint: When you replace a variable by a particular value, place parentheses around that value. Then use the order of operations to evaluate the expression.

Did you rewrite the expression as $3(-3)^2 - 7(-3) - 11$, using parentheses to complete the substitution?

Yes _____ No _____

If you answered No, please go back and perform that substitution step using parentheses around the specified value.

Did you next raise -3 to the second power to obtain $3(9) - 7(-3) - 11$?

Yes _____ No _____

If you answered No, review the order of operations and complete this step.

Remember that multiplication must be performed before addition and subtraction.

If you answered Problem 19 incorrectly, go back and rework this problem using these suggestions.

Simplifying Algebraic Expressions with Many Grouping Symbols—Problem 26

Simplify $-3\{a + b[3a - b(1-a)]\}$.

Helpful Hint: Work from the inside out. Remove the innermost symbols, $(\ )$, first. Then remove the next level of innermost symbols, $[\ ]$. Finally remove the outermost symbols, $\{\ \}$. Be careful to avoid sign errors.

Did you first obtain the expression $-3\{a + b[3a - b + ab]\}$ when you removed the innermost parentheses?

Yes _____ No _____

If you answered No, go back to the original problem and use the distributive property to remove the innermost grouping symbol, the parentheses.

Did you next obtain the expression $-3\{a + 3ab - b^2 + ab^2\}$ when you removed the brackets?

Yes _____ No _____

If you answered No, use the distributive property to distribute b on the outside of the bracket, $[\ ]$, to each term inside the bracket.

Finally, use the distributive property in the last step to distribute the -3 across all of the terms inside the outermost brackets, $\{\ \}$. Be careful with the $+/-$ signs.

Now go back and rework the problem using these suggestions.

Chapter 2 Equations, Inequalities, and Applications
2.1 The Addition Principle of Equality

Vocabulary

Equation • solution • equivalent equations • solving the equation • identity

satisfies • checking • addition principle • opposite in sign • additive inverse

1. A number that is opposite in sign to another number is called its _____.

2. A(n) _____ is a statement in which the equals sign $(=)$ is used to indicate that two expressions are equal.

3. If the same value appears on both sides of the equals sign, the equation is called a(n) _____.

4. If the same number is added to both sides of an equation, the _____ states that the results on both sides are equal.

Example	**Student Practice**
1. Solve for x. $x+16=20$	**2.** Solve for x. $x+8=42$

Use the addition principle to add -16 to both sides and simplify.

$$x+16=20$$
$$x+16+(-16)=20+(-16)$$
$$x+0=4$$
$$x=4$$

Substitute the value found for x into the original equation to verify the solution.

$$x+16=20$$
$$4+16\overset{?}{=}20$$
$$20=20$$

The solution checks.

Vocabulary Answers: 1. additive inverse 2. equation 3. identity 4. addition principle

Example	Student Practice
3. Solve for x. $1.5 + 0.2 = 0.3 + x + 0.2$	**4.** Solve for x. $2.7 - 0.2 = 0.4 + x - 0.6$

3. Solve for x. $1.5 + 0.2 = 0.3 + x + 0.2$

Simplify by adding.

$$1.5 + 0.2 = 0.3 + x + 0.2$$
$$1.7 = x + 0.5$$

Add -0.5 to both sides.

$$1.7 + (-0.5) = x + 0.5 + (-0.5)$$

Simplify.

$$1.2 = x$$

The check is left to the student.

5. Is 10 the solution to the equation $-15 + 2 = x - 3$? If not, find the solution.

Substitute 10 for x in the equation.

$$-15 + 2 = x - 3$$

$$-15 + 2 \overset{?}{=} 10 - 3$$

$$13 \neq 7$$

The values are not equal. Thus, 10 is not the solution. Solve the original equation to find the solution. Start by simplifying.

$$-15 + 2 = x - 3$$
$$-13 = x - 3$$

Add 3 to both sides.

$$-13 = x - 3$$
$$-13 + 3 = x - 3 + 3$$
$$-10 = x$$

The check is left to the student.

4. Solve for x. $2.7 - 0.2 = 0.4 + x - 0.6$

6. Is 26 the solution to the equation $x - 12 = 28 - 4$? If not, find the solution.

Example	Student Practice

7. Find the value of x that satisfies the equation.

$$\frac{1}{5} + x = -\frac{1}{10} + \frac{1}{2}$$

To be combined, the fractions must have common denominators. Rewrite each fraction as an equivalent fraction with a denominator of 10 and simplify.

$$\frac{1}{5} \cdot \frac{2}{2} + x = -\frac{1}{10} + \frac{1}{2} \cdot \frac{5}{5}$$

$$\frac{2}{10} + x = -\frac{1}{10} + \frac{5}{10}$$

$$\frac{2}{10} + x = \frac{4}{10}$$

Add $-\frac{2}{10}$ to each side.

$$\frac{2}{10} + \left(-\frac{2}{10}\right) + x = \frac{4}{10} + \left(-\frac{2}{10}\right)$$

$$x = \frac{2}{10} = \frac{1}{5}$$

Check.

$$\frac{1}{5} + x = -\frac{1}{10} + \frac{1}{2}$$

$$\frac{1}{5} + \left(\frac{1}{5}\right) \overset{?}{=} -\frac{1}{10} + \frac{1}{2}$$

$$\frac{2}{5} \overset{?}{=} -\frac{1}{10} + \frac{5}{10}$$

$$\frac{2}{5} \overset{?}{=} \frac{4}{10}$$

$$\frac{2}{5} = \frac{2}{5}$$

The solution checks.

8. Find the value of x that satisfies the equation.

$$\frac{1}{3} + x = -\frac{1}{21} + \frac{1}{7}$$

Extra Practice

1. Solve for x. Check your answers.

$$14 = 8 + x$$

2. Solve for x. Check your answers.

$$1.7 + 4.2 + x = 19.23 - 9.8$$

3. Is 6 the solution to the equation $x + 7 = 14 - 19 + 6$? If it is not, find the solution.

4. Is 28 the solution to the equation $-17 + x - 4 = 12 - 8 + 3$? If it is not, find the solution.

Concept Check

Explain how you would check to verify whether $x = 3.8$ is the solution to $-1.3 + 1.6 + 3x = -6.7 + 4x + 3.2$.

Chapter 2 Equations, Inequalities, and Applications
2.2 The Multiplication Principle of Equality

Vocabulary
terminating decimal • multiplicative inverse • multiplication principle • division principle

1. If both sides of an equation are multiplied by the same nonzero number, the
    _____ states that the results on both sides are equal.

2. The _____ states that dividing both sides of an equation by the same nonzero
    number results in an equivalent equation.

3. For any nonzero number a, the _____ is $\dfrac{1}{a}$.

4. A _____ is a decimal with a definite number of digits.

Example	Student Practice
1. Solve for x. $\dfrac{1}{3}x = -15$	**2.** Solve for x. $-\dfrac{1}{7}x = -3$

We know that $(3)\left(\dfrac{1}{3}\right) = 1$. Multiply each
side of the equation by 3 to isolate x.

$$3\left(\dfrac{1}{3}x\right) = 3(-15)$$

$$\left(\dfrac{3}{1}\right)\left(\dfrac{1}{3}\right)x = -45$$

$$x = -45$$

Check.

$$\dfrac{1}{3}(-45) \overset{?}{=} -15$$

$$-15 = -15$$

The solution checks.

Vocabulary Answers: 1. multiplication principle 2. division principle 3. multiplicative inverse
4. terminating decimal

Example	Student Practice
3. Solve for x. $5x = 125$	**4.** Solve for x. $6x = 42$

Divide both sides by 5.

$$\frac{5x}{5} = \frac{125}{5}$$
$$x = 25$$

Check.

$$5x = 125$$
$$5(25) \overset{?}{=} 125$$
$$125 = 125$$

The solution checks.

5. Solve for x. $9x = 60$	**6.** Solve for x. $16x = 30$

Divide both sides by 9 and simplify.

$$9x = 60$$
$$\frac{9x}{9} = \frac{60}{9}$$
$$x = \frac{20}{3}$$

The check is left to the student.

7. Solve for x. $-3x = 48$	**8.** Solve for x. $-9x = 63$

Divide both sides by -3.

$$\frac{-3x}{-3} = \frac{48}{-3}$$
$$x = -16$$

The check is left to the student.

Example	Student Practice
9. Solve for x. $-x = -24$	**10.** Solve for x. $-x = 15$

9. Solve for x. $-x = -24$

Rewrite the equation. Note that $-1x$ is the same as $-x$.

$$-1x = -24$$

Divide both sides by the coefficient -1.

$$\frac{-1x}{-1} = \frac{-24}{-1}$$
$$x = 24$$

The check is left to the student.

---

11. Solve for x. $-78 = 5x - 8x$

Combine the like terms on the right side.

$$-78 = 5x - 8x$$
$$-78 = -3x$$

Divide both sides by the coefficient -3.

$$\frac{-78}{-3} = \frac{-3x}{-3}$$
$$26 = x$$

The check is left to the student.

12. Solve for x. $24 = 4x - 8x$

---

13. Solve for x. $31.2 = 6.0x - 0.8x$

Combine the like terms on the right side.

$$31.2 = 6.0x - 0.8x$$
$$31.2 = 5.2x$$

Divide both sides by 5.2.
$$\frac{31.2}{5.2} = \frac{5.2x}{5.2}$$
$$6 = x$$

14. Solve for x. $15.4 = 4.8x - 2.6x$

Extra Practice

1. Solve for x. Check your answers.
   $$-13 = -x$$

2. Solve for x. Check your answers.
   $$-5 = \frac{x}{5}$$

3. Is 7 the solution to the equation $\frac{3}{7}x = 3$? If it is not, find the correct solution.

4. Is 0.12 the solution to the equation $3x = -0.36$? If it is not, find the correct solution.

Concept Check

Explain how you would check to verify whether $x = 36\frac{2}{3}$ is the solution to $-22 = -\frac{3}{5}x$.

Chapter 2 Equations, Inequalities, and Applications
2.3 Using the Addition and Multiplication Principles Together

Vocabulary
distributive property   •   like terms   •   multiplication principle   •   addition principle

1. To solve an equation of the form $ax^2 + bx = c$, we must use both the addition principle and the _____.

2. If the variable appears on both sides of the equation, apply the _____ to the variable term to collect the variable terms on one side of the equation.

3. If an equation contains parentheses, first apply the _____ to remove the parentheses.

4. Where it is possible, first collect _____ on one or both sides of the equation.

Example	**Student Practice**
1. Solve for x. $5x + 3 = 18$	**2.** Solve for x. $6x - 8 = 4$

Example

1. Solve for x. $5x + 3 = 18$

Use the addition principle to add -3 to both sides of the equation and simplify.

$$5x + 3 = 18$$
$$5x + 3 + (-3) = 18 + (-3)$$
$$5x = 15$$

Use the division principle to divide both sides by 5 and simplify.

$$5x = 15$$
$$\frac{5x}{5} = \frac{15}{5}$$
$$x = 3$$

Check. $5(3) + 3 \overset{?}{=} 18$

$$15 + 3 \overset{?}{=} 18$$
$$18 = 18$$

Student Practice

2. Solve for x. $6x - 8 = 4$

Vocabulary Answers: 1. multiplication principle 2. addition principle 3. distributive property 4. like terms

Example	Student Practice

3. Solve for x and check your solution.
$$9x+3=7x-2$$

Add $7x$ to both sides of the equation to get all variable terms on one side and combine like terms.

$$9x+3=7x-2$$
$$9x+(-7x)+3=7x+(-7x)-2$$
$$2x+3=-2$$

Add -3 to both sides and simplify.

$$2x+3=-2$$
$$2x+3+(-3)=-2+(-3)$$
$$2x=-5$$

Divide both sides by 2 and simplify.

$$2x=-5$$
$$\frac{2x}{2}=-\frac{5}{2}$$
$$x=-\frac{5}{2}$$

Check.

$$9x+3=7x-2$$
$$9\left(-\frac{5}{2}\right)+3\overset{?}{=}7\left(-\frac{5}{2}\right)-2$$
$$-\frac{45}{2}+3\overset{?}{=}-\frac{35}{2}-2$$
$$-\frac{45}{2}+\frac{6}{2}\overset{?}{=}-\frac{35}{2}-\frac{4}{2}$$
$$-\frac{39}{2}=-\frac{39}{2}$$

The solution checks.

4. Solve for x and check your solution.
$$4x+6=x+4$$

72

Example	Student Practice
5. Solve for x and check your solution. $5x + 26 - 6 = 9x + 12x = 25$	**6.** Solve for x and check your solution. $3x + 22 - 5 = 8x - x$

Combine like terms. Then add $(-5x)$ to both sides.

$$5x + 26 - 6 = 9x + 12x$$
$$5x + 20 = 21x$$
$$5x + (-5x) + 20 = 21x + (-5x)$$
$$20 = 16x$$

Divide both sides by 16.

$$\frac{20}{16} = \frac{16x}{16}$$
$$\frac{5}{4} = x$$

The check is left for the student.

Example	Student Practice
7. Solve for x and check your solution. $4(x+1) - 3(x-3) = 25$	**8.** Solve for x and check your solution. $-6(x-3) = 4(2x+5)$

Multiply by 4 and -3 to remove the parentheses. Then combine like terms. Be careful of the signs.

$$4(x+1) - 3(x-3) = 25$$
$$4x + 4 - 3x + 9 = 25$$
$$x + 13 = 25$$

Subtract 13 from both sides to isolate the variable.

$$x + 13 = 25$$
$$x + 13 - 13 = 25 - 13$$
$$x = 12$$

The check is left to the student.

Example	Student Practice
9. Solve for x and check your solution. $$0.3(1.2x - 3.6) = 4.2x - 16.44$$	**10.** Solve for z and check your solution. $$0.6(z - 2) = 0.4(z + 7)$$

Remove the parentheses.

$$0.3(1.2x - 3.6) = 4.2x - 16.44$$
$$0.36x - 1.08 = 4.2x - 16.44$$

Solve.

$$0.36x - 1.08 = 4.2x - 16.44$$
$$0.36x + (-0.36x) - 1.08 = 4.2x + (-0.36x) - 16.44$$
$$-1.08 + 16.44 = 3.84x - 16.44 + 16.44$$
$$15.32 = 3.84x$$
$$\frac{15.36}{3.84} = \frac{3.84x}{3.84}$$
$$4 = x$$

The check is left to the student.

Extra Practice

1. Solve for x and check your solution.
$$3x + 15 = 30$$

2. Solve for x and check your solution.
$$2x - 3 = x + 9$$

3. Solve for x and check your solution.
$$18 - 3 = 16x - 4x + 12$$

4. Solve for z and check your solution.
$$3(2z - 3) - 3 = 3z - 2(2z - 1)$$

Concept Check

Explain how you would solve the equation $3(x - 2) + 2 = 2(x - 4)$.

Chapter 2 Equations, Inequalities, and Applications
2.4 Solving Equations with Fractions

Vocabulary

LCD • multiplication principle • infinite number of solutions • no solution • decimal

1. A(n) _____ is a fraction written a special way.

2. If there is no value of x for which an equation is true, the equation has _____.

3. To eliminate the fractions in an equation, multiply each term by the _____.

Example	Student Practice
1. Solve for x. $\dfrac{1}{4}x - \dfrac{2}{3} = \dfrac{5}{12}x$	2. Solve for x. $\dfrac{1}{18}x + \dfrac{2}{3} = \dfrac{5}{6}x$

To eliminate the fractions, multiply both sides by the LCD, 12, and apply the distributive property to simplify.

$$\frac{1}{4}x - \frac{2}{3} = \frac{5}{12}x$$

$$12\left(\frac{1}{4}x - \frac{2}{3}\right) = 12\left(\frac{5}{12}x\right)$$

$$\left(\frac{12}{1}\right)\left(\frac{1}{4}\right)x - \left(\frac{12}{1}\right)\left(\frac{2}{3}\right) = \left(\frac{12}{1}\right)\left(\frac{5}{12}\right)x$$

$$3x - 8 = 5x$$

Add $-3x$ to both sides, then divide by 2.

$$3x - 8 = 5x$$

$$3x + (-3x) - 8 = 5x + (-3x)$$

$$-8 = 2x$$

$$-\frac{8}{2} = \frac{2x}{2}$$

$$-4 = x$$

The check is left to the student.

Vocabulary Answers: 1. decimal 2. no solution 3. LCD

Example	Student Practice
3. Solve for x and check your solution.	**4.** Solve for x and check your solution.

$$\frac{x}{3} + 3 = \frac{x}{5} - \frac{1}{3}$$

$$\frac{x}{2} + 2 = \frac{x}{3} - \frac{1}{2}$$

Multiply each term by the LCD, 15.

$$15\left(\frac{x}{3}\right) + 15(3) = 15\left(\frac{x}{5}\right) - 15\left(\frac{1}{3}\right)$$

$$5x + 45 = 3x - 5$$

Subtract $3x$ and 45 from both sides.

$$5x + 45 = 3x - 5$$
$$5x - 3x + 45 - 45 = 3x - 3x - 5 - 45$$
$$2x = -50$$

Divide by 2.

$$\frac{2x}{2} = \frac{-50}{2}$$
$$x = -25$$

Check.

$$\frac{(-25)}{3} + 3 \overset{?}{=} \frac{(-25)}{5} - \frac{1}{3}$$

$$-\frac{25}{3} + \frac{9}{3} \overset{?}{=} -\frac{5}{1} - \frac{1}{3}$$

$$-\frac{16}{3} \overset{?}{=} -\frac{15}{3} - \frac{1}{3}$$

$$-\frac{16}{3} = -\frac{16}{3}$$

The solution checks.

Example	Student Practice
5. Solve for x and check your solution.	**6.** Solve for x and check your solution.

5. Solve for x and check your solution.

$$\frac{1}{3}(x-2)=\frac{1}{5}(x+4)+2$$

Remove the parentheses.

$$\frac{x}{3}-\frac{2}{3}=\frac{x}{5}+\frac{4}{5}+2$$

Multiply all terms by the LCD, 15.
Then, solve the resulting equation.

$$15\left(\frac{x}{3}\right)-15\left(\frac{2}{3}\right)=15\left(\frac{x}{5}\right)+15\left(\frac{4}{5}\right)+15(2)$$

$$5x-10=3x+12+30$$
$$5x-10=3x+42$$
$$2x=52$$
$$x=26$$

The check is left to the student.

6. Solve for x and check your solution.

$$\frac{1}{4}(x-4)=\frac{1}{6}(x+2)+8$$

---

7. Solve for x.

$$0.2(1-8x)+1.1=-5(0.4x-0.3)$$

Remove parentheses. Then, multiply each term by 10 to get integer coefficients and solve.

$$0.2-1.6x+1.1=-2.0x+1.5$$
$$2-16x+11=-20x+15$$
$$4x+13=15$$
$$4x=2$$
$$x=\frac{1}{2} \text{ or } 0.5$$

The decimal form of the solution should only be given if it is a terminating decimal. The check is left to the student.

8. Solve for x.

$$0.3(1-6x)+1.3=3(-0.8x+0.6)$$

Extra Practice

1. Solve for p. Check your solution.

$$\frac{1}{3}p = p - 1$$

2. Solve for x. Check your solution.

$$\frac{x+2}{3} = \frac{x}{18} + \frac{1}{9}$$

3. Solve for x. Check your solution.

$$\frac{3}{4}(2x+6) - 3 = 3(x+3)$$

4. Solve for x. Check your solution.

$$0.4(x-6) = 0.7(2x+3) + 0.5$$

Concept Check

Explain how you would solve $\dfrac{x+5}{6} = \dfrac{x}{2} + \dfrac{3}{4}$.

Chapter 2 Equations, Inequalities, and Applications
2.5 Translating English Phrases into Algebraic Expressions

Vocabulary

more than • sum of • increased by • added to • greater than • plus • minus
decreased by • less than • subtracted from • smaller than • fewer than • of
diminished by • difference between • double • twice • product of • times
divided by • quotient of

1. "The _____ a number and three" translates to $x+3$.

2. "The _____ a number and four" translates to $x-4$.

3. "The _____ a two and a number" translates to $2x$.

4. "The _____ a number and five" translates to $\dfrac{x}{5}$.

Example	Student Practice
1. Write each English phrase as an algebraic expression.	**2.** Write each English phrase as an algebraic expression.
(a) A quantity is increased by five.	**(a)** Five less than a quantity
$x+5$	
(b) Double the value.	**(b)** Triple the discount.
$2x$	
(c) One-third of the weight	**(c)** One-fifth of the height
$\dfrac{x}{3}$ or $\dfrac{1}{3}x$	
(d) Seven less than a number	**(d)** Ten more than a number
$x-7$ Note that the variable or expression that follows the words "less than" always comes first.	

Vocabulary Answers: 1. sum of 2. difference between 3. product of 4. quotient of

Example	Student Practice
3. Write each English phrase as an algebraic expression.	**4.** Write each English phrase as an algebraic expression.
(a) Seven more than double a number	**(a)** Eight more than triple a number
$2x+7$	
(b) The value of the number is increased by seven and then doubled.	**(b)** The value of the number is increased by eight and then tripled.
Note that the word "then" tells us to add x and 7 before doubling.	
$2(x+7)$	
Note that this is not the same as $2x+7$.	**(c)** One-fourth of the sum of a number and 6
(c) One-half of the sum of a number and 3	
$\dfrac{1}{2}(x+3)$	
5. Use a variable and an algebraic expression to describe two quantities in the English sentence "Mike's salary is $2000 more than Fred's salary."	**6.** Use a variable and an algebraic expression to describe two quantities in the English sentence "Tom's car has 12,500 more miles on it than Chuck's truck."
The two quantities that are being compared are Mike's and Fred's salaries. Since Mike's salary is being compared to Fred's salary, we let the variable represent Fred's salary. The choice of the letter f helps us to remember that the variable represents Fred's salary.	
Let $f =$ Fred's salary.	
Then, $f + \$2000 =$ Mike's salary, since Mike's salary is $2000 more than Fred's.	

Example	Student Practice
7. The length of a rectangle is 3 meters shorter than twice the width. Use a variable and an algebraic expression to describe the length and the width. Draw a picture of the rectangle and label the length and width.	**8.** The length of a rectangle is 7 inches shorter than triple the width. Use a variable and an algebraic expression to describe the length and the width. Draw a picture of the rectangle and label the length and width.

7. The length of the rectangle is being compared to the width. Use the letter w for width. Let w = the width. Express the length in terms of the width. The length is 3 meters shorter than twice the width. Then $2w - 3$ = the length. A picture of the rectangle is shown.

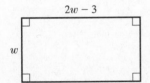

9. The first angle of a triangle is triple the second angle. The third angle of the triangle is 12° more than the second angle. Describe each angle algebraically. Draw a diagram of the triangle and label its parts.

Since the first and third angles are described in terms of the second angle, we let the variable represent the number of degrees in the second angle.

Let s = the number of degrees in the second angle. Then $3s$ = the number of degrees in the first angle and $s + 12$ = the number of degrees in the third angle.

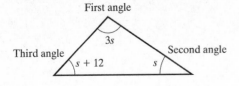

10. The first angle of a triangle is five times the second angle. The third angle of the triangle is 40° more than the second angle. Describe each angle algebraically. Draw a diagram of the triangle and label its parts.

Extra Practice

1. Write an algebraic expression for the quantity. Let x represent the unknown value.

   a value increased by 12

2. Write an algebraic expression for the quantity. Let x represent the unknown value.

   one-fourth of a number, decreased by one

3. Write an algebraic expression for the quantities being compared.

   The value of Alicia's car is $2300 more than the value of Allison's car.

4. Write an algebraic expression for the quantities being compared.

   Patricia owns 17 more comic books than Scott, Walter owns three times as many as Scott, and Adrienne owns five fewer than four times as many as Scott.

Concept Check

Explain how you would decide whether to use $\frac{1}{3}(x+7)$ or $\frac{1}{3}x+7$ as an algebraic expression for the phrase "one-third of the sum of a number and seven."

Chapter 2 Equations, Inequalities, and Applications
2.6 Using Equations to Solve Word Problems

Vocabulary
understand the problem • write an equation • solve and state the answer • check

1. When solving word problems, the last step is to _____, which involves checking the solution in the original equation and determining if the answer is reasonable.

2. When solving word problems, the third step is to _____, which involves solving the equation to determine the answer to the problem.

3. When solving word problems, the second step is to _____, which involves looking for key words to help you to translate the words into algebraic symbols and expressions.

Example	Student Practice
1. Two-thirds of a number is eighty-four. What is the number?	**2.** Four-fifths of a number is sixty-eight. What is the number?

1. Two-thirds of a number is eighty-four. What is the number?

Understand the problem. Draw a sketch. Let $x =$ the unknown number.

Write an equation. The word "of" translates to multiplication and the word "is" translates to equals.

$$\frac{2}{3}x = 84$$

Solve and state the answer.

$$\frac{2}{3}x = 84$$

$$2x = 252$$

$$x = 126$$

The check is left to the student.

2. Four-fifths of a number is sixty-eight. What is the number?

Vocabulary Answers: 1. check 2. solve and state the answer 3. write an equation

Example	Student Practice

3. Five more than six times a quantity is three hundred five. Find the number.

Understand the problem. Read the problem carefully. You may not need to draw a sketch. Let $x =$ the unknown quantity.

Write an equation. "More than" translates to addition, "times" translates to multiplication, and "is" translates to equals.

$$5 + 6x = 305$$

Solve and state the answer.

$$6x + 5 = 305$$
$$6x + 5 - 5 = 305 - 5$$
$$6x = 300$$
$$\frac{6x}{6} = \frac{300}{6}$$
$$x = 50$$

The quantity, or number, is 50.

Check. Is five more than six times 50 three hundred five?

$$6(50) + 5 \overset{?}{=} 305$$
$$300 + 5 \overset{?}{=} 305$$
$$305 = 305$$

The answer checks.

4. Nineteen more than four times a quantity is two hundred seventy-nine. Find the number.

Example	Student Practice

5. The smallest angle of an isosceles triangle measures $24°$. The other two angles are larger. What are the measurements of the other two angles?

6. The largest angle of an isosceles triangle measures $116°$. The other two angles are smaller. What are the measurements of the other two angles?

Let $x =$ the measure in degrees of each of the larger angles. Draw a sketch.

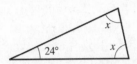

Write an equation showing the sum of all three angles is $180°$, solve for x.

$$24° + x + x = 180°$$
$$x = 78°$$

---

7. Two people travel in separate cars. They each travel a distance of 330 miles on an interstate highway. To maximize fuel economy, Fred travels at exactly 50 mph. Sam travels at exactly 55 mph. How much time did the trip take each person?

8. Two people travel in separate cars. They each travel a distance of 780 miles on an interstate highway. To maximize fuel economy, John travels at exactly 65 mph. Yuri travels at exactly 60 mph. How much time did the trip take each person?

Read the problem carefully and create a Mathematics Blueprint. Write an equation using the formula
$$\text{distance} = (\text{rate})(\text{time}) \text{ or } d = rt.$$
Substitute the known values into the formula and solve for t.

$$d = rt \qquad\qquad d = rt$$
$$330 = 50t_f \text{ and } 330 = 55t_s$$
$$6.6 = t_f \qquad\qquad 6.6 = 6t_s$$

It took Fred 6.6 hours. It took Sam 6 hours. The check is left to the student.

Extra Practice

1. What number minus 312 gives 234? Check your solution.

2. A number is tripled and then increased by 27. The result is 72. What is the original number? Check your solution.

3. The local health food store has six times as many energy drinks as energy bars. There are 108 energy drinks in stock. How many energy bars are in stock? Check to see if your answer is reasonable.

4. Amy's cell phone company charges $10.50 per month for 200 minutes of use and $0.15 for each additional minute. Last month Amy's cell phone bill was $55.50. How many additional minutes was she charged for? Check to see if your answer is reasonable.

Concept Check

Explain how you would set up an equation to solve the following problem.

Phil purchased two shirts for $23 each and then purchased several pairs of socks. The socks were priced at $0.75 per pair. How many pairs of socks did he purchase if the total cost was $60.25?

Chapter 2 Equations, Inequalities, and Applications
2.7 Solving Word Problems: The Value of Money and Percents

Vocabulary

percent   •   simple interest   •   compound interest   •   interest

1. _____ is a charge for borrowing money or an income from investing money.

2. _____ is computed by multiplying the amount of money borrowed or invested times the rate of interest times the period of time over which it is borrowed or invested.

3. Applied situations often require finding a _____ of an unknown number.

Example	Student Practice
1. A business executive rented a car. The Supreme Car Rental Agency charged $39 per day and $0.28 per mile. The executive rented the car for two days and the total rental cost was computed to be $176. How many miles did the executive drive the rented car? Understand the problem. It is known that it costs $176 to rent the car for two days. It is necessary to find the number of miles the car was driven. Let $m =$ the number of miles driven in the rented car. Write an equation. Use the relationship for calculating the total cost. per-day cost + mileage cost = total cost $(39)(2) + (0.28)m = 176$ Solve and state the answer. $78 + 0.28m = 176$ $0.28m = 98$ $m = 350$ The executive drove 350 miles. The check is left to the student.	**2.** A business woman rented a room at a motel. The motel charged $52 per day and $2.50 per hour of internet use. The business woman stayed in the room for four days and the total charge was computed to be $248. How many hours of internet use did the business woman accrue?

Vocabulary Answers: 1. interest 2. simple interest 3. percent

Example	Student Practice
3. A sofa was marked with the following sign: "The price of this sofa has been reduced by 23%. You save $138 if you buy now." What was the original price of the sofa?	**4.** A refrigerator was marked with the following sign: "The price of this refrigerator has been reduced by 33%. You save $264 if you buy now." What was the original price of the refrigerator?

Understand the problem. Let $s =$ the original price of the sofa. Then $0.23s =$ the amount of the price reduction, which is $138.

Write an equation and solve.

$$0.23s = 138$$

$$\frac{0.23s}{0.23} = \frac{138}{0.23}$$

$$s = 600$$

The original price of the sofa was $600. The check is left to the student.

Example	Student Practice
5. A woman invested an amount of money in two accounts for one year. She invested some at 8% simple interest and the rest at 6% simple interest. Her total amount invested was $1250. At the end of the year she had earned $86 in interest. How much money had she invested in each account?	**6.** A man invested an amount of money in two accounts for one year. He invested some at 6% simple interest and the rest at 5% simple interest. His total amount invested was $3500. At the end of the year he had earned $195 in interest. How much money had he invested in each account?

The simple interest formula is $I = prt$.

Let $x =$ amount invested at 8%. Then $\$1250 - x =$ amount invested at 6%.

Write an equation and solve for x.

$$0.08x + 0.06(1250 - x) = 86$$

$$x = 550$$

The amount invested at 8% is $550. The amount invested at 6% is $700.

Example	Student Practice

7. When Bob got out of math class, he had to make a long-distance call. He had exactly enough dimes and quarters to make a phone call that would cost $2.55. He had one fewer quarter than he had dimes. How many coins of each type did he have?

Let d = the number of dimes. Then $d-1$ = the number of quarters. The total value of the coins was $2.55. Each dime is worth $0.10 and each quarter is worth $0.25. So, d dimes are worth $0.10d$ dollars and $(d-1)$ quarters are worth $0.25(d-1)$ dollars. Now write an equation for the total value, and solve.

$$0.10d + 0.25(d-1) = 2.55$$
$$0.10d + 0.25d - 0.25 = 2.55$$
$$0.35d - 0.25 = 2.55$$
$$0.35d = 2.80$$
$$d = 8$$

Find the number of quarters Bob has, $d-1 = 8-1 = 7$. Thus, Bob has eight dimes and seven quarters.

Check the answer. Bob has $8-7 = 1$ less quarter than he has dimes. Check that eight dimes and seven quarters are worth $2.55.

$$8(\$0.10) + 7(\$0.25) \overset{?}{=} \$2.55$$
$$\$0.80 + \$1.75 \overset{?}{=} \$2.55$$
$$\$2.55 = \$2.55$$

The solution checks.

8. When Rebecca got out of chemistry class, she went to the vending machines for a snack. She had exactly enough nickels and dimes to get a combination of items costing $2.75. She had four fewer dimes than she had nickels. How many coins of each type did she have?

Extra Practice

1. Angelina received a pay raise this year. The raise was 5% of last year's salary. This year, Angelina earned $17,640. What was her salary before the raise?

2. Find the simple interest on $6,500 borrowed at 15% for one year.

3. Randall is due to receive a 7% raise, which in dollars will be $3,780 per year. What is his current salary?

4. Mr. Finch keeps money in his pillowcase. Right now, he has equal numbers of five, ten, and twenty-dollar bills, with no other denominations. He has exactly $1505. How many bills does he have all together?

Concept Check

Explain how you would set up an equation to solve the following problem.

Robert has $2.55 in change consisting of nickels, dimes, and quarters. He has twice as many dimes as quarters. He has one more nickel than he has quarters. How many of each coin does he have?

Chapter 2 Equations, Inequalities and Applications
2.8 Solving Inequalities in One Variable

Vocabulary

inequalities • is less than • is greater than • solution • solution set • graph

1. The set of all numbers that make the inequality true is called a(n) _____.

2. Comparisons of values, such as one value being greater than or less than another value, are called _____.

3. A(n) _____ is a visual representation of the solution set.

4. One number _____ another if it is to the right of the other on the number line.

Example	Student Practice
1. In each statement, replace the question mark with the symbol < or >.	**2.** In each statement, replace the question mark with the symbol < or >.
(a) 3 ? −1	**(a)** −4 ? −10
$3 > -1$ because 3 is to the right of −1 on the number line.	
(b) −2 ? 1	**(b)** 2 ? −2
$-2 < 1$ because −2 is to the left of 1.	
(c) −3 ? −4	**(c)** −1 ? 4
$-3 > -4$ because −3 is to the right of −4.	
(d) 0 ? 3	**(d)** 0 ? −7
$0 < 3$ because 0 is to the left of 3.	
(e) −3 ? 0	**(e)** 5 ? 8
$-3 < 0$ because −3 is left of 0.	

Vocabulary Answers: 1. solution set 2. inequality 3. graph 4. is greater than

Example	Student Practice
3. State each mathematical relationship in words and then graph it.	**4.** State each mathematical relationship in words and then graph it.

3.

(a) $x < -2$

We state "x is less than -2."

$x < -2$

(b) $-3 < x$

We can state that "-3 is less than x" or, equivalently, that "x is greater than -3." Be sure you see that $-3 < x$ is equivalent to $x > -3$. Although both statements are correct, we usually write the variable first in a simple inequality containing a variable and a numerical value.

4.

(a) $x < 3$

(b) $\dfrac{5}{2} \geq x$

5. Translate each English sentence into an algebraic statement.

(a) The police on the scene said that the car was traveling more than 80 miles per hour. (Use the variable s for speed.)

Since the speed must be greater than 80, we have $s > 80$.

(b) The owner of the trucking company said that the payload of a truck must never exceed 4500 pounds. (Use the variable p for payload.)

If the payload of the truck can never exceed 4500 pounds, then the payload must always be less than or equal to 4500 pounds. Thus we write $p \leq 4500$.

6. Translate each English sentence into an algebraic statement.

(a) A student's budget is very tight, so when shopping for a car, her car payment must be less than 125 dollars per month.

(b) The maximum number of people allowed on the boat at one time is 35.

Example	**Student Practice**

7. Solve and graph. $3x+7 \geq 13$

Subtract 7 from both sides.

$$3x+7 \geq 13$$
$$3x+7-7 \geq 13-7$$
$$3x \geq 6$$

Divide both sides by 3.

$$3x \geq 6$$
$$\frac{3x}{3} \geq \frac{6}{3}$$
$$x \geq 2$$

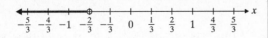

8. Solve and graph. $6x-4 > 5$

---

9. Solve and graph. $5-3x > 7$

Subtract 5 from both sides and simplify.

$$5-5-3x > 7-5$$
$$-3x > 2$$

Divide by -3. When dividing by a negative number, the inequality is reversed.

$$-3x > 2$$
$$\frac{-3x}{-3} < \frac{2}{-3}$$
$$x < -\frac{2}{3}$$

Note the direction of the inequality. The graph is as follows.

10. Solve and graph. $8-3x < 9$

Example	Student Practice
11. Solve and graph. $-\dfrac{13x}{2} \le \dfrac{x}{2} - \dfrac{15}{8}$	**12.** Solve and graph. $-\dfrac{7x}{3} < \dfrac{x}{3} + \dfrac{23}{27}$

Multiply by the LCD, 8, and simplify.

$$8\left(-\frac{13x}{2}\right) \le 8\left(\frac{x}{2}\right) - 8\left(\frac{15}{8}\right)$$
$$-52x \le 4x - 15$$

Subtract both $4x$ from both sides and combine like terms. Then, solve the inequality. Be sure to reverse the direction of the inequality symbol.

$$-52x - 4x \le 4x - 4x - 15$$
$$-56x \le -15$$
$$x \ge \frac{15}{56}$$

The graph is as follows.

Extra Practice

1. In the statement, replace the question mark with the symbol $<$ or $>$.

$$\frac{1}{3} \ ? \ 5$$

2. Graph. $x < -4$

3. Translate the English sentence into an algebraic statement.

1,000,000 is greater than or equal to the total weight w.

4. Solve and graph. $3x + 5 < 14$

Concept Check

Explain the difference between $12 < x$ and $x < 12$. Would the graphs of these inequalities be the same or different?

MATH COACH

Mastering the skills you need to do well on the test.

Watch the **MATH COACH** videos in MyMathLab® or on YouTube™ while you work the problems below. These helpful hints will help you avoid making common errors on test problems.

Solving Equations with Both Parentheses and Decimals—

Problem 6 Solve for the variable. $0.8x + 0.18 - 0.4x = 0.3(x + 0.2)$

Helpful Hint: After removing parentheses, it might be most helpful for you to multiply both sides of the equation by 100 in order to obtain a simpler, equivalent equation without decimals. Check to make sure that you did not make any errors in calculations before solving the equation.

Did you remove the parentheses to get the equation $0.8x + 0.18 - 0.4x = 0.3x + 0.06$?
Yes ____ No ____

If you answered No, go back and use the distributive property to remove the parentheses. Be careful to place the decimal point in the correct location when multiplying 0.3 and 0.2 together.

Did you multiply each term of the equation by 100 to move the decimal point two places to the right to get the equivalent equation $80x + 18 - 40x = 30x + 6$?
Yes ____ No ____

If you answered No, stop and carefully complete this step before solving the equation. Remember that you may need to

add a 0 to the end of a term in order to move the decimal point two places to the right.

If you answered Problem 6 incorrectly, go back and rework the problem using these suggestions.

Solving Equations with More Than One Set of Parentheses—Problem 12

Solve for the variable. $20 - (2x + 6) = 5(2 - x) + 2x$

Helpful Hint: Slowly complete the necessary steps to remove each set of parentheses before doing any other steps. Be careful to avoid sign errors.

Did you obtain the equation $20 - 2x - 6 = 10 - 5x + 2x$ after removing each set of parentheses?
Yes ____ No ____

If you answered No, go back and carefully use the distributive property to remove each set of parentheses. Locate any mistakes you have made and make a note of the type of error discovered.

Did you combine like terms to get the equation $14 - 2x = 10 - 3x$?
Yes ____ No ____

If you answered No, stop and perform that step correctly.

Now go back and rework the problem using these suggestions.

Solving Equations with Both Fractions and Parentheses—Problem 17

Solve for x. $\frac{2}{3}(x+8)+\frac{3}{5}=\frac{1}{5}(11-6x)$

Helpful Hint: Remove the parentheses first. This is the most likely place to make a mistake. Next, carefully show every step of your work as you multiply each fraction by the LCD. Be sure to check your work.

Did you remove each set of parentheses to obtain the equation $\frac{2}{3}x+\frac{16}{3}+\frac{3}{5}=\frac{11}{5}-\frac{6}{5}x$?

Yes _____ No _____

If you answered No, stop and carefully redo your steps of multiplication, showing every part of your work.

Did you identify the LCD as 15 and then multiply each term by 15 to get $10x+80+9=33-18x$?
Yes _____ No _____

If you answered No, stop and write out your steps slowly.

If you answered Problem 17 incorrectly, go back and rework the problem using these suggestions.

Solving and Graphing Inequalities on a Number Line—Problem 19

Solve and graph the inequality. $2-7(x+1)-5(x+2)<0$

Helpful Hint: Be sure to remove parentheses and combine any like terms on each side of the inequality before solving for the variable. Always verify the following:
1) Did you multiply or divide by a negative number? If so, did you reverse the inequality symbol?
2) In the graph, is your choice of an open circle or closed circle correct?

Did you remove parentheses to get $2-7x-7-5x-10<0$?
Did you combine like terms to obtain the inequality $-15-12x<0$? Next, did you add 15 to both sides of the inequality? Yes _____ No _____

If you answered No to any of these questions, stop now and perform those steps.

Did you remember to reverse the inequality symbol in the last step? Yes _____ No _____

If you answered No, please review the rules for when to reverse the inequality symbol and then go back and perform this step.

Did you use an open circle in your number line graph? Is your arrow pointing to the right? Yes _____ No _____

If you answered No to either question, please review the rules for how to graph an inequality involving the < or > inequality symbols.

Now go back and rework the problem using these suggestions.

Name: _____   Date: _____

Instructor: _____   Section: _____

Chapter 3 Graphing and Functions
3.1 The Rectangular Coordinate System

Vocabulary

graphs • rectangular coordinate system • origin • x-axis • y-axis
ordered pair • coordinates • x-coordinate • y-coordinate • solution

1. The numbers in an ordered pair are often referred to as the _____ of the point.

2. We can illustrate algebraic relationships with drawings called _____.

3. The vertical number line above the origin is often called the _____.

4. The first number in an ordered pair is called the _____ and it represents the distance from the origin measured along the horizontal axis.

Example	Student Practice
1. Plot the point $(5,2)$ on a rectangular coordinate system. Label this point as *A*.	**2.** Plot the point $(2,5)$ on the preceding rectangular coordinate system. Label this point as *B*.

Since the x-coordinate is 5, we first count 5 units to the right on the x-axis. Then, because the y-coordinate is 2, we count 2 units up from the point where we stopped on the x-axis. This locates the point corresponding to $(5,2)$. We mark this point with a dot and label it *A*.

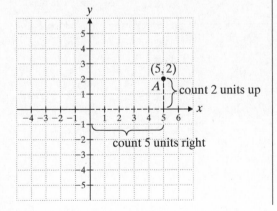

Vocabulary Answers: 1. coordinates 2. graphs 3. y-axis 4. x-coordinate

Example	Student Practice

3. Use the rectangular coordinate system to plot each point. Label the points F and G respectively.

(a) $(-5, -3)$

Notice that the x-coordinate, -5, is negative. On the coordinate grid, negative x-values appear to the left of the origin. Thus we will begin by counting 5 squares to the left, starting at the origin. Since the y-coordinate, -3, is negative, we will count 3 units down from the point where we stopped on the x-axis.

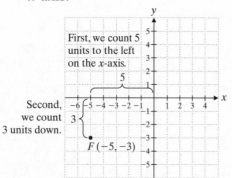

(b) $(2, -6)$

The x-coordinate is positive. Begin by counting 2 squares to the right of the origin. Then count down because the y-coordinate is negative.

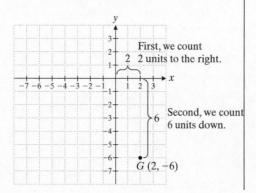

4. Use the rectangular coordinate system below to plot each point. Label the points I, J, and K, respectively.

(a) $(-1, -4)$

(b) $(-4, 3)$

(c) $(3, -5)$

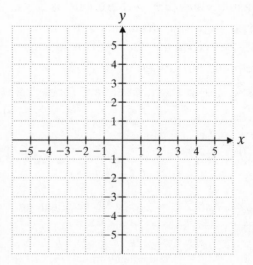

Example	Student Practice

5. What ordered pairs of numbers represent point A and point B on the graph?

To find point A, move along the x-axis until you get as close as possible to A, ending at 5. Thus obtaining 5 as the first number of the ordered pair. Then count 4 units upward on a line parallel to the y-axis to reach A. So you obtain 4 as the second number of the ordered pair. Thus point A is $(5,4)$. Use the same approach to find point B: $(-5,-3)$.

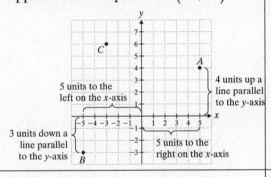

6. What ordered pair of numbers represents point C on the graph in Example **5**?

7. Is the ordered pair $(-1,4)$ a solution to the equation $3x + 2y = 5$?

We replace the values for x and y to see if we obtain a true statement.
Replace x with -1 and y with 4.

$$3x + 2y = 5$$
$$3(-1) + 2(4) \overset{?}{=} 5$$
$$-3 + 8 \overset{?}{=} 5$$
$$5 = 5$$

The ordered pair $(-1,4)$ is a solution to $3x + 2y = 5$ because when we replace x with -1 and y with 4, we obtain a true statement.

8. Is the ordered pair $(3,-3)$ a solution to the equation $3x + 2y = 5$?

Example	Student Practice
9. Find the missing coordinate to complete the ordered-pair solution $(0,?)$ to the equation $2x+3y=15$.	**10.** Find the missing coordinate to complete the ordered-pair solution $(?,3)$ to the equation $6x+5y=3$.

For the ordered pair $(0,?)$, we know that $x=0$. Replace x with 0 in the equation and solve for y.

$$2x+3y=15$$
$$2(0)+3y=15$$
$$0+3y=15$$
$$y=5$$

Thus we have the ordered pair $(0,5)$.

Extra Practice

1. Plot the point $D:(-1,-3)$.

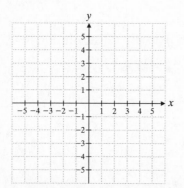

2. Consider the point plotted on the graph below. Give the coordinates for point A.

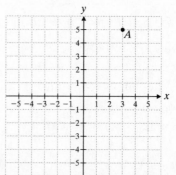

3. Find the missing coordinate to complete the ordered-pair solution to $y=-2x+3$.

  (a) $(-1,?)$

  (b) $(3,?)$

4. Find the missing coordinate to complete the ordered-pair solution to $4x-2y=12$.

  (a) $(-1,?)$

  (b) $(4,?)$

Concept Check
Explain how you would find the missing coordinate to complete the ordered-pair solution to the equation $2.5x+3y=12$ if the ordered pair was of the form $(?,-6)$.

Chapter 3 Graphing and Functions
3.2 Graphing Linear Equations

Vocabulary
linear equation   •   x-intercept   •   y-intercept   •   horizontal   •   vertical

1. The _____ of a line is the point where the line crosses the y-axis.

2. The graph of any _____ in two variables is a straight line.

3. The _____ of a line is the point where the line crosses the x-axis.

Example	Student Practice
1. Find three ordered pairs that satisfy $y = -2x + 4$. Then graph the resulting straight line.	**2.** Find three ordered pairs that satisfy $y = 2x - 3$. Then graph the resulting straight line. Use the given coordinate system.

Example:

Let $x = 0$, $x = 1$, and $x = 3$. For each x-value, find the corresponding y-value in the equation. Place each y-value in the table next to its x-value.

x	y
0	4
1	2
3	-2

If we plot these ordered pairs and connect the three points, we get a straight line that is the graph of the equation.

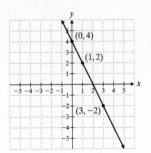

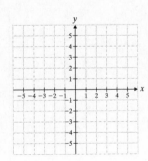

Vocabulary Answers: 1. y-intercept   2. linear equation   3. x-intercept

Example	Student Practice

3. Graph $5x - 4y + 2 = 2$.

First, we simplify the equation by subtracting 2 from each side.

$$5x - 4y + 2 = 2$$
$$5x - 4y + 2 - 2 = 2 - 2$$
$$5x - 4y = 0$$

Since we are free to choose any value of x, $x = 0$ is a natural choice. Calculate the value of y when $x = 0$.

$$5(0) - 4y = 0$$
$$-4y = 0$$
$$y = 0$$

A convenient choice for a replacement of x is a number that is divisible by 4. Let $x = 4$ and $x = -4$. Follow the process used above to find the corresponding y-values and place the results in the table of values.

x	y
0	0
4	5
−4	−5

Graph the line.

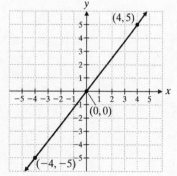

4. Graph $4x - 2y + 7 = 7$.

Example	Student Practice

Example

5. Complete **(a)** and **(b)** for the equation $5y - 3x = 15$.

(a) State the x- and y-intercepts.

Let $y = 0$.

$$5(0) - 3x = 15$$
$$-3x = 15$$
$$x = -5$$

$(-5, 0)$ is the x-intercept. Now let $x = 0$.

$$5y - 3(0) = 15$$
$$5y = 15$$
$$y = 3$$

$(0, 3)$ is the y-intercept.

(b) Use the intercept method to graph.

Find a third point and then graph. If we let $y = 6$, $x = 5$. The ordered pair is $(5, 6)$.

x	y
−5	0
0	3
5	6

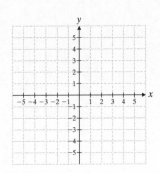

Student Practice

6. Complete **(a)** and **(b)** for the equation $y + 2x = -2$.

(a) State the x- and y-intercepts.

(b) Use the intercept method to graph.

Example	Student Practice
7. Graph $2x + 1 = 11$.	**8.** Graph $y + 4 = 7$.

Notice that there is only one variable, x, in the equation. Simplifying the equation yields $x = 5$. Since the x-coordinate of every point on this line is 5, we can see that the vertical line will be 5 units to the right of the y-axis.

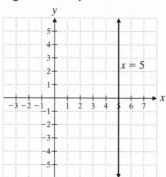

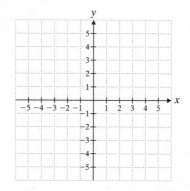

Extra Practice

1. Graph $y = 2x - 5$ by plotting three points and connecting them.

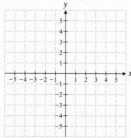

2. Graph $5x + 2y = 10$ by plotting three points and connecting them.

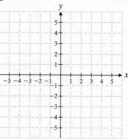

3. Graph $y = 5 - x$ by plotting intercepts and one other point.

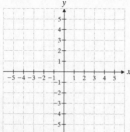

4. Graph $2x - 5 = y$ by plotting intercepts and one other point.

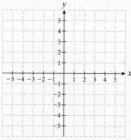

Concept Check

In graphing the equation $3y - 7x = 0$, what is the most important ordered pair to obtain before drawing a graph of the line? Why is that ordered pair so essential to drawing the graph?

Chapter 3 Graphing and Functions
3.3 The Slope of a Line

Vocabulary

slope   •   positive slope   •   negative slope   •   zero slope
undefined slope   •   slope-intercept form

1. In a coordinate plane, the _____ of a straight line is defined by the change in y divided by the change in x.

2. A vertical line is said to have _____.

3. If we know the slope and the y-intercept, we can write the equation of the line in _____.

Example	Student Practice
1. Find the slope of the line that passes through $(2,0)$ and $(4,2)$. Let $(2,0)$ be the first point (x_1, y_1) and $(4,2)$ be the second point (x_2, y_2). $$\text{slope} = m = \frac{y_2 - y_1}{x_2 - x_1} = \frac{2-0}{4-2} = \frac{2}{2} = 1$$ Note that the slope of the line will be the same if we let $(4,2)$ be the first point (x_1, y_1) and $(2,0)$ be the second point (x_2, y_2). $$m = \frac{y_2 - y_1}{x_2 - x_1} = \frac{0-2}{2-4} = \frac{-2}{-2} = 1$$ Thus, it does not matter which point you call (x_1, y_1) and which you call (x_2, y_2).	**2.** Find the slope of the line that passes through $(0,-4)$ and $(-3,-6)$.

Vocabulary Answers: 1. slope 2. undefined slope 3. slope-intercept form

Example	Student Practice
3. Find the slope of the line that passes through the given points. **(a)** $(0,2)$ and $(5,2)$ Calculate the slope. $$m = \frac{y_2 - y_1}{x_2 - x_1} = \frac{2-2}{5-0} = \frac{0}{5} = 0$$ The slope of a horizontal line is 0. **(b)** $(-4,0)$ and $(-4,-4)$ Calculate the slope. $$m = \frac{y_2 - y_1}{x_2 - x_1} = \frac{-4-0}{-4-(-4)} = \frac{-4}{0}$$ Recall that division by 0 is undefined. The slope of a vertical line is undefined.	**4.** Find the slope of the line that passes through the given points. **(a)** $(7,3)$ and $(7,-4)$ **(b)** $(5,3)$ and $(-1,3)$
5. What is the slope and the y-interept of the line $5x + 3y = 2$? We want to solve for y and get the equation in the form $y = mx + b$. $5x + 3y = 2$ $\qquad 3y = -5x + 2$ $\qquad y = -\frac{5}{3}x + \frac{2}{3}$ $m = -\frac{5}{3}$ and $b = \frac{2}{3}$ The slope is $-\frac{5}{3}$. The y-interept is $\left(0, \frac{2}{3}\right)$.	**6.** What is the slope and the y-interept of the line $9x + 3y = 12$?

Example	Student Practice
7. Find an equation of the line with slope $\frac{2}{5}$ and y-interept $(0,-3)$.	**8.** Find an equation of the line with slope $\frac{3}{4}$ and y-interept $(0,-7)$.
(a) Write the equation in slope-intercept form, $y = mx + b$.	**(a)** Write the equation in slope-intercept form, $y = mx + b$.
We are given that $m = \frac{2}{5}$ and $b = -3$. Thus we have the following.	

$$y = mx + b$$
$$y = \frac{2}{5}x + (-3)$$
$$y = \frac{2}{5}x - 3$$

(b) Write the equation in the form $Ax + By = C$.	**(b)** Write the equation in the form $Ax + By = C$.
Clear the equation of fractions so that A, B, and C are integers.	

$$y = \frac{2}{5}x - 3$$
$$5y = 5\left(\frac{2x}{5}\right) - 5(3)$$
$$5y = 2x - 15$$

Subtract $2x$ from each side. Then, multiply each term by -1, because the form $Ax + By = C$ is usually written with A as a positive integer.

$$5y = 2x - 15$$
$$-2x + 5y = -15$$
$$2x - 5y = 15$$

Example	Student Practice
9. Graph the equation $y = -\dfrac{1}{2}x + 4$.	**10.** Graph the equation $y = -\dfrac{3}{4}x + 1$.

Begin with the y-intercept. Since $b = 4$, plot the point $(0,4)$. The slope, $-\dfrac{1}{2}$ can be written as $\dfrac{-1}{2}$. Begin at $(0,4)$ and go down 1 unit and to the right 2 units. This is the point $(2,3)$. Plot the point. Draw the line that connects the two points. This is the graph of the equation $y = -\dfrac{1}{2}x + 4$.

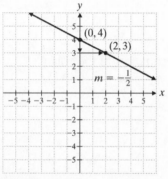

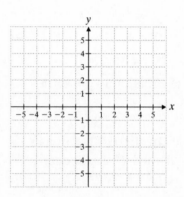

Extra Practice

1. Find the slope of a straight line that passes through the points $(-5,-2)$ and $(1,-4)$.

2. Find the slope and the y-intercept of the line $y = 5x$.

3. Write the equation of the line in slope-intercept form given $m = -4$ and the y-intercept is $\left(0, \dfrac{4}{5}\right)$.

4. Write the equation of the line in slope-intercept form given $m = 0$ and the y-intercept is $(0,-2)$.

Concept Check

Consider the formula for slope: $m = \dfrac{y_2 - y_1}{x_2 - x_1}$. Explain why we substitute the y coordinates in the numerator.

108

Chapter 3 Graphing and Functions
3.4 Writing the Equation of a Line

Vocabulary

slope   • y-intercept   •   slope-intercept form   •   vertical units   •   horizontal units

parallel lines   •   perpendicular lines

1. Given the slope and a point on the line we can find the _____.

2. To find the slope of a line given the graph, we count the number of _____ and horizontal units from one point on the line to another.

3. _____ have slopes whose product is −1.

4. _____ have the same slope but different y-intercepts.

Example	**Student Practice**
1. Find an equation of the line that passes through $(-3, 6)$ with slope $-\dfrac{2}{3}$.	**2.** Find an equation of the line that passes through $(2, -5)$ with slope $-\dfrac{1}{2}$.

Example (continued)

We are given the values $m = -\dfrac{2}{3}$, $x = -3$, and $y = 6$. Substitute the given values of x, y, and m into the equation $y = mx + b$. Solve for b.

$$y = mx + b$$

$$6 = \left(-\frac{2}{3}\right)(-3) + b$$

$$6 = 2 + b$$

$$4 = b$$

Use the values of b and m to write the equation in the form $y = mx + b$.

An equation of the line is $y = -\dfrac{2}{3}x + 4$.

Vocabulary Answers: 1. y-intercept 2. vertical units 3. perpendicular lines 4. parallel lines

Example	Student Practice
3. Find an equation of the line that passes through $(2,5)$ and $(6,3)$.	**4.** Find an equation of the line that passes through $(5,4)$ and $(10,1)$.

3. Find an equation of the line that passes through $(2,5)$ and $(6,3)$.

We first find the slope of the line. Then proceed as in Example **1**.

Substitute $(x_1, y_1) = (2,5)$ and $(x_2, y_2) = (6,3)$ into the formula.

$$m = \frac{y_2 - y_1}{x_2 - x_1}$$

$$m = \frac{y_2 - y_1}{x_2 - x_1} = \frac{3-5}{6-2} = \frac{-2}{4} = -\frac{1}{2}$$

Choose either point, say $(2,5)$, to substitute into $y = mx + b$ as in Example **1**. Then solve for b.

$$y = mx + b$$
$$5 = -\frac{1}{2}(2) + b$$
$$5 = -1 + b$$
$$6 = b$$

Use the values for b and m to write the equation.

An equation of the line is $y = -\frac{1}{2}x + 6$.

Note: We could have substituted the slope and the other point, $(6,3)$, into the slope-intercept form and arrived at the same answer. Try it.

Example	Student Practice

5. What is the equation of the line in the figure below?

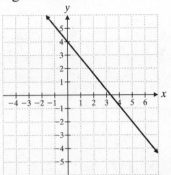

First, look for the y-intercept. The line crosses the y-axis at $(0,4)$. Thus $b = 4$.

Second, find the slope.

$$m = \frac{\text{change in } y}{\text{change in } x}$$

Look for another point on the line. We choose $(5,-2)$. Count the number of vertical units from 4 to -2 (rise). Count the number of horizontal units from 0 to 5 (run).

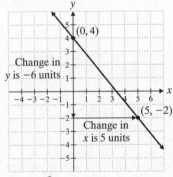

$$m = \frac{-6}{5}$$

Now, using $m = \dfrac{-6}{5}$ and $b = 4$, we can write an equation of the line.

$$y = mx + b$$

$$y = -\frac{6}{5}x + 4$$

6. What is the equation of the line in the figure below?

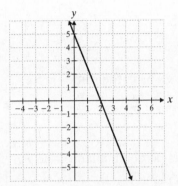

Example	Student Practice

7. Line h has a slope of $-\dfrac{2}{3}$.

   (a) If line f is parallel to line h, what is its slope?

   Parallel lines have the same slope.

   Line f has a slope of $-\dfrac{2}{3}$.

   (b) If line g is perpendicular to line h, what is its slope?

   Perpendicular lines have slopes whose product is -1.

$$m_1 m_2 = -1$$

$$-\frac{2}{3}m_2 = -1$$

$$m_2 = \frac{3}{2}$$

   Line g has a slope of $\dfrac{3}{2}$.

8. Line h has a slope of $\dfrac{5}{7}$.

   (a) If line f is parallel to line h, what is its slope?

   (b) If line g is perpendicular to line h, what is its slope?

Extra Practice

1. Find the equation of the line that passes through $(3,1)$ and has slope $-\dfrac{1}{2}$.

2. Write an equation of the line that passes through $(3,4)$ and $(-1,-16)$.

3. Write an equation of the line that passes through $\left(1,\dfrac{1}{6}\right)$ and $\left(2,\dfrac{4}{3}\right)$.

4. Find the equation of a line that passes through $(3,-7)$ and is parallel to $y = -5x + 2$.

Concept Check

How would you find an equation of the line that passes through $(-2,-3)$ and has zero slope?

Name: _____   Date: _____
Instructor: _____   Section: _____

Chapter 3 Graphing and Functions
3.5 Graphing Linear Inequalities

Vocabulary
linear inequality   •   solution   •   solid line   •   dashed line   •   test point

1. The _____ of an inequality is the set of all possible ordered pairs that when substituted into the inequality will yield a true statement.

2. If the _____ is a solution of the inequality, we shade the region on the side of the line that includes the point.

3. We use a _____ to indicate that the points on the line are included in the solution of the inequality.

Example	**Student Practice**
1. Graph $5x+3y>15$.	**2.** Graph $4x+3y>12$.

1. Graph $5x+3y>15$.

Begin by graphing the line $5x+3y=15$. Since there is no equals sign in the inequality, draw a dashed line to indicate that the line is not part of the solution set. The easiest test point to test is $(0,0)$. Substitute $(0,0)$ for (x,y).

$$5x+3y>15$$
$$5(0)+3(0)>15$$
$$0>15 \quad \text{false}$$

$(0,0)$ is not a solution. Shade the region on the side of the line that does not include $(0,0)$.

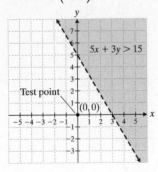

Vocabulary Answers: 1. solution 2. test point 3. solid line

Example	Student Practice
3. Graph $2y \le -3x$.	**4.** Graph $4y \le -5x$.

First, graph $2y = -3x$. Since $\le$ is used, the line will be a solid line.

We see that the line passes through $(0,0)$. We choose another test point. We will choose $(-3,-3)$.

$$2y \le -3x$$
$$2(-3) \le -3(-3)$$
$$-6 \le 9 \quad \text{true}$$

Since $(-3,-3)$ is a solution to the inequality, shade the region that includes $(-3,-3)$, that is the region below the line.

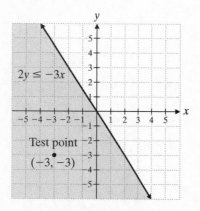

Example	Student Practice

5. Graph $x < -2$.

First, graph $x = -2$. Since $<$ is used, the line will be dashed.

Second, test $(0, 0)$ in the inequality.

$x < -2$

$0 < -2$ false

Since $(0, 0)$ is not a solution to the inequality, shade the region that does not include $(0, 0)$, that is the region to the left of the line $x = -2$.

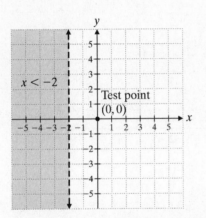

Observe that every point in the shaded region has an x-value that is less than -2.

6. Graph $x > -3$.

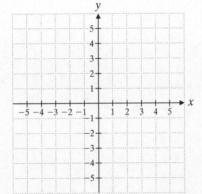

Extra Practice

1. Graph $y < 2x + 1$.

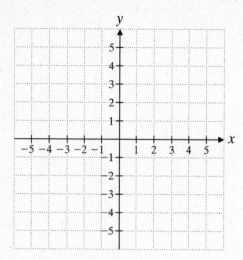

2. Graph $3x - 5y - 10 \geq 0$.

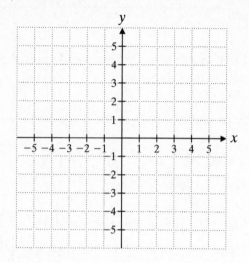

3. Graph $y \leq 4$.

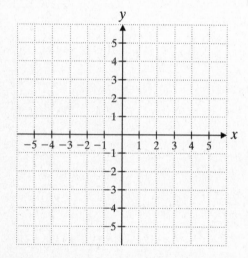

4. Graph $3x + 6y - 9 < 0$.

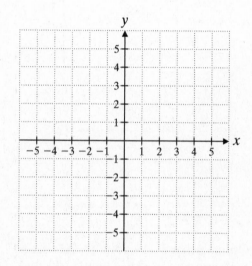

Concept Check

Explain how you would determine if you should shade the region above the line or below the line if you were to graph the inequality $y > -3x + 4$ using $(0,0)$ as a test point.

Chapter 3 Graphing and Functions
3.6 Functions

Vocabulary

independent variable • dependent variable • relation • domain • range
function • absolute zero • vertical line test • function notation

1. A(n) _____ is any set of ordered pairs.

2. A(n) _____ is a relation in which no two different ordered pairs have the same
   first coordinate.

3. The _____ is used to determine whether a relation is a function.

4. All the first coordinates in all of the ordered pairs of the relation make up the
   _____ of the relation.

Example	**Student Practice**
1. State the domain and range of the relation. $\{(5,7),(9,11),(10,7),(12,14)\}$	**2.** State the domain and range of the relation. $\{(-3,6),(7,1),(-2,1),(4,6)\}$
The domain consists of all the first coordinates in the ordered pairs. The first coordinates are 5, 9, 10, and 12.	
The range consists of all the second coordinates in the ordered pairs. The second coordinates are 7, 11, 7, and 14.	
We usually list the values of a domain or range from smallest to largest.	
The domain is $\{5,9,10,12\}$.	
The range is $\{7,11,14\}$.	
Note that we list 7 only once.	

Vocabulary Answers: 1. relation 2. function 3. vertical line test 4. domain

Example	Student Practice
3. Determine whether the relation is a function. **(a)** $\{(3,9),(4,16),(5,9),(6,36)\}$ No two ordered pairs have the same first coordinate. Thus this set of ordered pairs defines a function. **(b)** $\{(7,8),(9,10),(12,13),(7,14)\}$ Two different ordered pairs, $(7,8)$ and $(7,14)$ have the same first coordinate. Thus this relation is not a function.	**4.** Determine whether the relation is a function. **(a)** $\{(9,3),(16,4),(9,5),(36,6)\}$ **(b)** $\{(-3,17),(2,1),(4,-3),(7,17)\}$
5. Graph $y = x^2$. Begin by constructing a table of values. We select values for x and then determine by the equation the corresponding values of y. We then plot the ordered pairs and connect the points with a smooth curve.	**6.** Graph $x = y^2$. 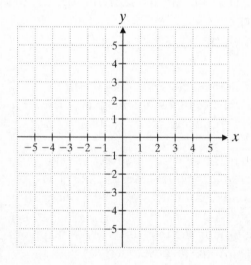

x	$y = x^2$	y
-2	$y = (-2)^2 = 4$	4
-1	$y = (-1)^2 = 1$	1
0	$y = (0)^2 = 0$	0
1	$y = (1)^2 = 1$	1
2	$y = (2)^2 = 4$	4

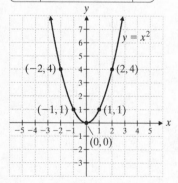

Example	**Student Practice**

7. Determine whether each of the following is the graph of a function.

(a)

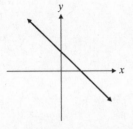

The graph of a straight line is a function. Any vertical line will cross this straight line in only one location.

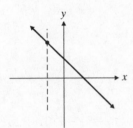

(b)

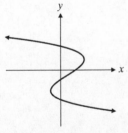

This graph is not the graph of a function. There exists a vertical line that will cross the curve in more than one place.

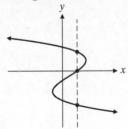

8. Determine whether each of the following is the graph of a function.

(a)

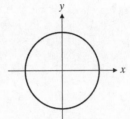

(b)

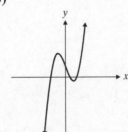

Example	Student Practice
9. If $f(x) = 3x^2 - 4x + 5$, find each of the following.	**10.** If $f(x) = 4x^2 - 2x + 7$, find each of the following.

9. (a) $f(-2)$

$$f(-2) = 3(-2)^2 - 4(-2) + 5$$
$$= 3(4) + -4(-2) + 5$$
$$= 12 + 8 + 5 = 25$$

(b) $f(4)$

$$f(4) = 3(4)^2 - 4(4) + 5$$
$$= 3(16) + -4(4) + 5$$
$$= 48 - 16 + 5 = 37$$

(c) $f(0)$

$$f(0) = 3(0)^2 - 4(0) + 5 = 5$$

10. (a) $f(-3)$

(b) $f(2)$

(c) $f(0)$

Extra Practice

1. Find the domain and range of the relation. Determine whether the relation is a function.

$$\{(2.5,3),(3.5,0),(5.5,-2),(8.5,-6)\}$$

2. Given $f(x) = 2x^2 - 3x + 1$, find the indicated values.
   (a) $f(0)$
   (b) $f(-3)$
   (c) $f(3)$

3. Determine whether the relation is a function.

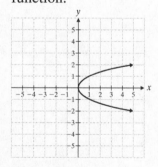

4. Graph $y = -3x^2$.

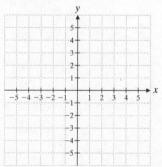

Concept Check

In the relation $\{(3,4),(5,6),(3,8),(2,9)\}$, why is there a different number of elements in the domain than the range?

MATH COACH

Mastering the skills you need to do well on the test.

Watch the MATH COACH videos in MyMathLab®or on You Tube™ while you work the problems below. These helpful hints will help you avoid making common errors on test problems.

Graphing a Linear Equation by Plotting Three Ordered Pairs—Problem 4   Graph $y = \frac{2}{3}x - 4$.

Helpful Hint: Find three ordered pairs that are solutions to the equation. Plot those 3 points. Then draw a line through the points. When the equation is solved for y and there are fractional coefficients on x, it is sometimes a good idea to choose values for x that result in integer values for y. This will make graphing easier.

Choosing values for x that result in integer values for y means choosing 0 or multiples of 3 since 3 is the denominator of the fractional coefficient on x. When we multiply by these values, the result becomes an integer.

Did you choose values for x that result in values for y that are not fractions? Yes _____ No _____

If you answered No, try using $x = 0$, $x = 3$, and $x = 6$. Solve the equation for y in each case to find the y-coordinate. Remember that it will make the graphing process easier if you choose values for x that will clear the fraction from the equation.

Did you plot the three points and connect them with a line? Yes _____ No _____

If you answered No, go back and complete this step.

If you feel more comfortable using the slope-intercept method to solve this problem, simply identify the y-intercept from the equation in $y = mx + b$ form, and then use the slope m to find two other points.

If you answered Problem 4 incorrectly, go back and rework the problem using these suggestions.

Write the Equation of a Line Given Two Points—Problem 10

Find an equation for the line passing through $(5, -4)$ and $(-3, 8)$.

Helpful Hint: When given two points (x_1, y_1) and (x_2, y_2), we can find the slope m using the slope formula $m = \frac{y_2 - y_1}{x_2 - x_1}$. Then you can substitute m into the equation $y = mx + b$ along with the coordinates of one of the points to find b, the y-intercept.

When you substituted the points into the slope formula to find m, did you obtain either $m = \frac{8 - (-4)}{-3 - 5}$ or

$m = \frac{-4 - 8}{5 - (-3)}$?

Yes _____ No _____

If you answered No, check your work to make sure you substituted the points correctly. Be careful to avoid any sign errors.

(continued on next page)

Did you use $m = -\dfrac{3}{2}$ and either of the points given when you substituted the values into the equation $y = mx + b$ to find the value of b? Yes _____ No _____

If you answered No, stop and make a careful substitution for $m = -\dfrac{3}{2}$ and either $x = 5$ and $y = -4$ or $x = -3$ and $y = 8$.

See if you can solve the resulting equation for b.

Now go back and rework the problem using these suggestions.

Graphing Linear Inequalities in Two Variables—Problem 12

Graph the region described by $-3x - 2y > 10$.

> **Helpful Hint:** First graph the equation $-3x - 2y = 10$. Determine if the line should be solid or dashed. Then pick a test point to see if it satisfies the inequality $-3x - 2y > 10$. If the test point satisfies the inequality, shade the side of the line on which the point lies. If the test point does not satisfy the inequality, shade the opposite side of the line.

Examine your work. Does the line $-3x - 2y = 10$ pass through the point $(0, -5)$? Yes _____ No _____

If you answered No, substitute $x = 0$ into the equation and solve for y. Check the calculations for each of the points you plotted to find the graph of this equation.

Did you draw a solid line? Yes _____ No _____

If you answered Yes, look at the inequality symbol. Remember that we only use a solid line with the symbols $\leq$ and $\geq$. A dashed line is used for $<$ and $>$.

Did you shade the area above the dashed line? Yes _____ No _____

If you answered Yes, stop now and use $(0, 0)$ as a test point and substitute it into the inequality $-3x - 2y > 10$. Then use the Helpful Hint to determine which side to shade.

If you answered Problem 12 incorrectly, go back and rework the problem using these suggestions.

Using Function Notation to Evaluate a Function—Problem 16(a) and 16(b)

For $f(x) = -x^2 - 2x - 3$:   **(a)** find $f(0)$.   **(b)** find $f(-2)$.

> **Helpful Hint:** Replace x with the number indicated. It is a good idea to place parentheses around the value to avoid any sign errors. Then use the order of operations to evaluate the function in each case.

(a) Did you replace x with 0 and write

$$f(0) = -(0)^2 - 2(0) - 3?$$

Yes _____ No _____

If you answered No, take time to go over your steps one more time, remembering that 0 times any number is 0.

(b) Did you replace x with -2 and write

$$f(0) = -(0)^2 - 2(0) - 3?$$

Yes _____ No _____

If you answered No, go over your steps again, remembering to place parentheses around -2.

Note that $(-2)^2 = 4$ and therefore

$$-(-2)^2 = -4.$$

Now go back and rework the problem again using these suggestions.

Chapter 4 Systems of Equations
4.1 Solving a System of Equations in Two Variables by Graphing

Vocabulary

system of equations • consistent • inconsistent • dependent • coordinate system
slope • linear equation • intersection

1. A linear system of equations that has one solution is said to be _____.

2. A(n) _____, is any set of equations in multiple variables that is considered at the same time.

3. A system of equations with infinite solutions is called _____.

4. A system of equations that has no solution is said to be _____.

Example	**Student Practice**
1. Solve by graphing. $\begin{array}{l} -2x + 3y = 6 \\ 2x + 3y = 18 \end{array}$	**2.** Solve by graphing. $\begin{array}{l} 2x + 2y = 4 \\ -3x + y = -2 \end{array}$

Solve each equation for y. Use a table to find and graph a series of ordered pairs.

$$y = \frac{2}{3}x + 2$$

$$y = -\frac{2}{3}x + 6$$

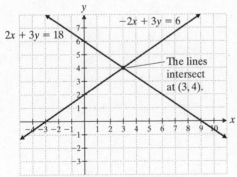

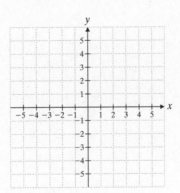

The unique solution is the point of intersection, $(3, 4)$.

Vocabulary Answers: 1. consistent 2. system of equations 3. dependent 4. inconsistent

Example	Student Practice

3. Solve by graphing. $3x - y = 1$
$3x - y = -7$

4. Solve by graphing. $2x + 4y = 0$
$\frac{1}{2}x + y = 3$

Graph both equations on the same coordinate system.

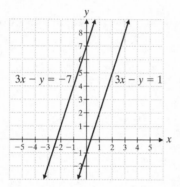

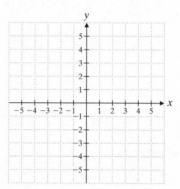

The two lines are parallel. A system of linear equations that does not intersect does not have a solutions and is called inconsistent.

5. Solve by graphing. $x + y = 4$
$3x + 3y = 12$

6. Solve by graphing. $-3x + 6y = 24$
$-x + 2y = 8$

Graph both equations on the same coordinate system.

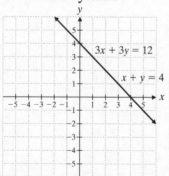

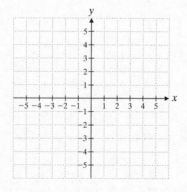

The two equations represent the same line; they coincide. There is an infinite number of solutions to this system. Such equations are said to be dependent.

Example	Student Practice

7. Roberts Plumbing and Heating charges $40 for a house call and then $35 per hour for labor. Instant Plumbing Repairs charges $70 for a house call and then $25 per hour for labor.

(a) Create a cost equation for each company, where y is the total cost of plumbing repairs and x is the number of hours per labor. Write a system of equations.

For each company obtain a cost equation.

Cost of plumbing $=$ *house call* $+$ *per hour* $\times$ *labor hours*

$y = 40 + 35 \times x$

$y = 70 + 25 \times x$

(b) Graph the two equations and determine from your graph how many hours of plumbing repairs would be required for the two companies to charge the same.

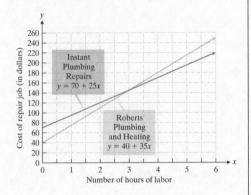

We see that the graphs of the two lines intersect at $(3, 145)$. Thus the two companies will charge the same if 3 hours of plumbing repairs are required.

8. Suppose Walter and Barbara try two more plumbers. Plumbers A charges $50 for a house call and then $40 per hour for labor. Plumbers B charges $90 for a house call and then $30 per hour for labor.

(a) Create a cost equation for each company, where y is the total cost of plumbing repairs and x is the number of hours per labor. Write a system of equations.

(b) Graph the two equations.

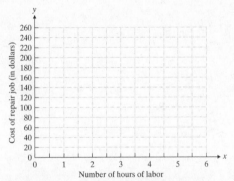

(c) Determine from your graph how many hours of plumbing repairs would be required for the two companies to charge the same.

(d) Determine from your graph which company charges less if the estimated amount of time to complete the plumbing repairs is 5 hours.

Extra Practice

1. Solve by graphing. If there isn't a unique solution to the system, state the reason.
$$3x - y = 0$$
$$2x + y = -10$$

2. Solve by graphing. If there isn't a unique solution to the system, state the reason.
$$-6x + 2y = 5$$
$$3x - y = 3$$

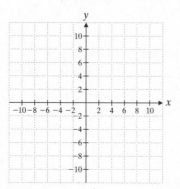

3. Solve by graphing. If there isn't a unique solution to the system, state the reason.
$$2x + 3y = 18$$
$$3x - 2y = 14$$

4. Solve by graphing. If there isn't a unique solution to the system, state the reason.
$$4x - 8y = 20$$
$$-2x + 4y = -10$$

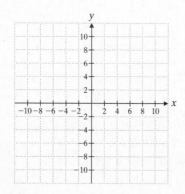
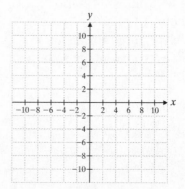

Concept Check

You are attempting to find the solution to this system of equations by graphing.

$$-3x + 4y = -16$$
$$0 = 6x - 8y + 16$$

What will you discover?

Chapter 4 Systems of Equations
4.2 Solving a System of Equations in Two Variables by the Substitution Method

Vocabulary
substitution method   •   system of equations   •   consistent   •   inconsistent
linear equation   •   dependent

1. An equation of the form $Ax + By = C$ is called a(n) _____.

2. The _____ is a strategy for solving a system of two linear equations that involves reducing the system to one equation in one variable which can be solved for.

Example

1. Find the solution. $x - 2y = 7$ (1)

$\qquad\qquad\qquad\qquad -5x + 4y = -5$ (2)

Solve equation (1) for x.

$x = 7 + 2y$

Substitute this expression into (2).

$$-5x + 4y = -5$$
$$-5(7 + 2y) + 4y = -5$$

Solve this equation for y.

$$-35 - 10y + 4y = -5$$
$$-6y = 30$$
$$y = -5$$

Use the value for y to find the value for x.

$x = 7 + 2y$

$x = 7 + 2(-5)$

$x = -3$

The solution is $(-3, -5)$.

Student Practice

2. Find the solution. $3x + y = -3$

$\qquad\qquad\qquad\qquad -5x - 2y = 4$

Vocabulary Answers: 1. linear equation 2. substitution method

Example	Student Practice
3. Find the solution. $\dfrac{3}{2}x + y = \dfrac{5}{2}$ (1)	**4.** Find the solution. $\dfrac{4}{3}x - \dfrac{2}{3}y = \dfrac{1}{12}$
$-y + 2x = -1$ (2)	$-y - x = -1$

Clear the first equation of fractions.

$$2\left(\dfrac{3}{2}x\right) + 2(y) = 2\left(\dfrac{5}{2}\right)$$

$$3x + 2y = 5$$

The new system is as follows.

$$3x + 2y = 5 \quad (3)$$
$$-y + 2x = -1 \quad (2)$$

Solve for y in equation (2).

$$y = 1 + 2x$$

Substitute this expression into (3) and solve for x.

$$3x + 2(1 + 2x) = 5$$
$$7x + 2 = 5$$
$$x = \dfrac{3}{7}$$

Use the value for x to find the value for y.

$$y = 1 + 2x$$
$$y = 1 + 2\left(\dfrac{3}{7}\right)$$
$$y = \dfrac{13}{7}$$

The solution is $\left(\dfrac{3}{7}, \dfrac{13}{7}\right)$.

128

Extra Practice

1. Find the solution to the system of equations by the substitution method. Write your solution in the form (a,b).

$$5a + 7b = -21$$
$$a + 18 = -6b$$

2. Find the solution to the system of equations by the substitution method. Write your solution in the form (x, y).

$$x = 6 + 4y$$
$$-3(x + y) = -36$$

3. Find the solution to the system of equations by the substitution method. Write your solution in the form (x, y).

$$8x - 5y = -78$$
$$6x - 3y = -54$$

4. Find the solution to the system of equations by the substitution method. Write your solution in the form (x, y).

$$5x + 8(y - 3) = 15$$
$$2(x + 2y) = 42$$

Concept Check

Explain how you would solve the following system by the substitution method.

$$3x - y = -4$$
$$-2x + 4y = 36$$

Chapter 4 Systems of Equations
4.3 Solving a System of Equations in Two Variables by the Addition Method

Vocabulary

substitution method   •   system of equations   •   addition method   •   elimination method

1. The addition method is often referred to as the _____.

2. The _____ is a strategy for solving a system of two linear equations that works well for integer, fractional, and decimal coefficients.

Example

1. Solve by addition. $5x + 2y = 7$ (1)

$$3x - y = 13 \quad (2)$$

Make the coefficients of the y-terms opposites by multiplying each term of equation (2) by 2.

$$2(3x) - 2(y) = 2(13)$$

Multiply, then add the equations.

$$5x + 2y = 7$$
$$\underline{6x - 2y = 26}$$
$$11x \quad\;\; = 33$$
$$x = 3$$

Substitute $x = 3$ into (1) and solve for y.

$$5(3) + 2y = 7$$
$$15 + 2y = 7$$
$$2y = -8$$
$$y = -4$$

The solution is $(3, -4)$.

Student Practice

2. Solve by addition. $5x - y = 21$

$$3x + 3y = 9$$

Vocabulary Answers: 1. elimination method 2. addition method

Example	Student Practice

3. Solve. $x - \dfrac{5}{2}y = \dfrac{5}{2}$ $\quad(1)$

$\qquad \dfrac{4}{3}x + y = \dfrac{23}{3}$ $\quad(2)$

4. Solve. $\dfrac{3}{4}x - \dfrac{5}{4}y = -\dfrac{9}{4}$

$\qquad \dfrac{1}{3}x + y = \dfrac{11}{3}$

Multiply each term of (1) by 2 and each term of (2) by 3 to clear the fractions.

$$2(x) - 2\left(\dfrac{5}{2}y\right) = 2\left(\dfrac{5}{2}\right)$$

$$3\left(\dfrac{4}{3}x\right) + 3(y) = 3\left(\dfrac{22}{3}\right)$$

$$2x - 5y = 5 \qquad (3)$$
$$4x + 3y = 23 \qquad (4)$$

Make the coefficients of the x-terms opposites by multiplying each term of equation (3) by -2. Add and solve for y.

$$-4x + 10y = -10$$
$$\underline{4x + 3y = 23}$$
$$13y = 13$$
$$y = 1$$

Substitute $y = 1$ into (4) and solve for x.

$$4x + 3(1) = 23$$
$$4x = 20$$
$$x = 5$$

The solution is $(5, 1)$.

The check is left to the student.

Example	Student Practice

5. Solve. $0.12x + 0.05y = -0.02$ (1)
           $0.08x - 0.03y = -0.14$ (2)
	6. Solve. $0.12x - 0.67y = -1.03$
	           $0.50x + 0.75y = -0.75$

Since the decimals are hundredths, multiply each term of both equations by 100.

$$100(0.12x) + 100(0.05y) = 100(-0.02)$$
$$100(0.08x) - 100(0.03y) = 100(-0.14)$$

$$12x + 5y = -2 \qquad (3)$$
$$8x - 3y = -14 \qquad (4)$$

Make the coefficients of the y-terms opposites by multiplying each term of (3) by 3 and each term of (4) by 5.

$$36x + 15y = -6$$
$$\underline{40x - 15y = -70}$$
$$76x = -76$$
$$x = -1$$

Substitute x into (3) and solve for y.

$$12(-1) + 5y = -2$$
$$y = 2$$

The solution is $(-1, 2)$.

To check, substitute the solution into the original equations.

$$0.12(-1) + 0.05(2) = -0.02$$
$$-0.02 = -0.02 \quad \text{Checks}$$
$$0.08(-1) - 0.03(2) = -0.14$$
$$-0.14 = -0.14 \quad \text{Checks}$$

Extra Practice

1. Find the solution by the addition method. Check your answers.

$$x - y = 6$$

$$7x - 9y = -4$$

2. Find the solution by the addition method. Check your answers.

$$7x + 13 = 9y$$

$$3x + 5y = -41$$

3. Find the solution by the addition method. Check your answers.

$$x + \frac{3}{2}y = 3$$

$$7x - 9y = -5$$

4. Find the solution by the addition method. Check your answers.

$$6x - y = 5$$

$$6x + 3y = -3$$

Concept Check

Explain how you would obtain an equivalent system that does not have decimals if you wanted to solve the following system.

$$0.5x + 0.4y = -3.4$$

$$0.02x + 0.03y = -0.08$$

Chapter 4 Systems of Equations
4.4 Review of Methods for Solving Systems of Equations

Vocabulary
substitution method • addition method • inconsistent • identity • dependent

1. The equation $7 = 7$ is always true; it is an example of a(n) _____.

2. The _____ works well if one or more variables have a coefficient of 1 or -1.

3. A(n) _____ system of linear equations is a system of parallel lines that are not equal.

4. _____ equations produce a system with an unlimited number of solutions.

Example

1. Select a method and solve the system of equations.

$$x + y = 3080$$
$$2x + 3y = 8740$$

Use substitution since there are x- and y-values that have coefficients of 1. Solve the first equation for y and substitute it into the other to find x.

$$y = 3080 - x$$
$$2x + 3(3080 - x) = 8740$$
$$2x + 9240 - 3x = 8740$$
$$x = 500$$

Substitute 500 for x and solve for y.

$$y = 3080 - 500$$
$$y = 2580$$

The solution is $(500, 2580)$.

Student Practice

2. Select a method and solve the system of equations.

$$3x - 4y = 13$$
$$9x - 2y = 119$$

Vocabulary Answers: 1. identity 2. substitution method 3. inconsistent 4. dependent

Example	**Student Practice**
3. Solve algebraically. $\begin{array}{l}3x - y = -1 \\ 3x - y = -7\end{array}$	**4.** Solve algebraically. $\begin{array}{l}-3x - y = -5 \\ 6x + 2y = -1\end{array}$

The addition method would be very convenient in this case.

Multiply each term in the second equation by -1 and then add the two equations.

$$3x - y = -1$$
$$\underline{-3x + y = +7}$$
$$0 = 6$$

The statement $0 = 6$ is not true, thus there is no solution to this system of equations. The lines are parallel and the system is inconsistent.

---

5. Solve algebraically. $\begin{array}{l}x + y = 4 \\ 3x + 3y = 12\end{array}$	**6.** Solve algebraically. $\begin{array}{l}4x - y = 3 \\ 12x - 3y = 9\end{array}$

Let us use the substitution method. Solve the first equation for y and substitute it into the second equation.

$$y = 4 - x$$
$$3x + 3(4 - x) = 12$$
$$3x + 12 - 3x = 12$$
$$12 = 12$$

$12 = 12$ is always true; it is an identity. This means all the solutions of one equation of the system are also solutions of the other equation. Therefore, the lines coincide and there are an infinite number of solutions to this system of equations.

Extra Practice

1. If possible, solve by an algebraic method. Otherwise, state that the problem has no solution or an infinite number of solutions.

$$9x + 1 = 7y$$
$$-2x - 4y = 28$$

2. If possible, solve by an algebraic method. Otherwise, state that the problem has no solution or an infinite number of solutions.

$$9x - 3y = 10$$
$$45x - 15y = 20$$

3. If possible, solve by an algebraic method. Otherwise, state that the problem has no solution or an infinite number of solutions.

$$\frac{3}{2}x + \frac{1}{2}y - 7 = 0$$
$$\frac{2}{5}x - \frac{3}{5}y + \frac{9}{5} = 0$$

4. If possible, solve by an algebraic method. Otherwise, state that the problem has no solution or an infinite number of solutions.

$$3x + y = -14$$
$$y = 3x + 4$$

Concept Check

Explain how you would remove the fractions in order to solve the following system.

$$\frac{3}{10}x + \frac{2}{5}y = \frac{1}{2}$$
$$\frac{1}{8}x - \frac{3}{16}y = -\frac{1}{2}$$

Chapter 4 Systems of Equations
4.5 Solving Word Problems Using Systems of Equations

Vocabulary
substitution method   •   addition method   •   elimination method   •   system of equations

1. When solving a word problem with two equations and two unknown it is often helpful to set up a(n) _____.

2. If one or more variables have a coefficient of 1 or −1, the _____ works well.

Example	Student Practice
1. A worker in a large post office is trying to verify the rate at which two electronic card-sorting machines operate. Yesterday the first machine sorted for 3 minutes and the second machine sorted for 4 minutes. The total workload both machines processed during that time was 10,300 cards. Two days ago the first machine sorted for 2 minutes and the second machine sorted for 3 minutes. The total workload both machines processed during that time period was 7400 cards. Can you determine the number of cards per minute sorted by each machine?	**2.** A landscape company is trying to verify the rate at which two employees can plant bulbs in a large park. Yesterday the first employee planted for 5 hours and the second employee planted for 2 hours. The total number of bulbs planted during that time was 115. Last week the first employee planted for 3 hours and the second employee planted for 4 hours. The total number of bulbs planted during that time was 125. Can you determine the number of bulbs planted per hour by each employee?

Let $x =$ the number of cards per minute sorted by the first machine and $y =$ the number of cards per minute sorted by the second machine and write a system of two equations with two unknowns using the information from the two days.

$3x + 4y = 10,300$ Yesterday

$2x + 3y = 7400$ Two days ago

Solve the system using the addition method. The first machine sorts 1300 cards per minute and the second machine sorts 1600 cards per minute.

Vocabulary Answers: 1. system of equations 2. substitution method

Example	Student Practice

3. A lab technician is required to prepare 200 liters of a solution. The prepared solution must contain 42% fungicide. The technician wishes to combine a solution that contains 30% fungicide with a solution that contains 50% fungicide. How much of each solution should he use?

We need to find the amount of each solution that will give us 200 liters of a 42% solution. Let $x =$ the number of liters of the 30% solution needed and $y =$ the number of liters of the 50% solution needed.

Write an equation for combining x and y to get 200 liters.

$x + y = 200$

Write an equation for the percent of fungicide in each solution (30% of x liters, 50% of y liters, and 42% of 200 liters).

$0.3x + 0.5y = 0.42(200) = 84$

Solve the following system of equations using the addition method.

$$x + y = 200$$
$$0.3x + 0.5y = 84$$

The technician should use 80 liters of the 30% fungicide and 120 liters of the 50% fungicide to obtain the required solution.

The check is left to the student.

4. A chemistry student is required to prepare 150 milliliters of a solution. The prepared solution must contain 15% alcohol. The student wishes to combine a solution that contains 25% alcohol with a solution that contains 10% alcohol. How much of each solution should he use?

Example	Student Practice
5. Mike recently rode his boat on Lazy River. He took a 48-mile trip up the river against the current in exactly 3 hours. He refueled and made the return trip in exactly 2 hours. What was the speed of his boat in still water and the speed of the current in the river?	**6.** Gary flew his small airplane 150 nautical miles to visit a friend. On the way there, he flew against a strong wind for 2.5 hours. On the way home, he flew in the same direction as the wind and made the trip in exactly 1.5 hours. What was the speed of his plane in still air and the speed of the wind that day?

Let b = the speed of the boat in still water in miles/hour and let c = the speed of the river current in miles/hour.

When we travel against the current, the current is slowing us down. Since the current's speed opposes the boat's speed in still water, we must subtract: $b - c$.

When we travel with the current, the current is helping us travel forward. The current's speed is added to the boat's speed in still water, we must add: $b + c$.

Use the distance formula,

$$distance = rate \times time$$

to write a system of equations.

$48 = (b - c) \cdot 3$ Against the current
$48 = (b + c) \cdot 2$ With the current

Remove the parenthesis to get the following system of equations.

$48 = 3b - 3c$ Against the current
$48 = 2b + 2c$ With the current

Solve the system of equations using the addition method to find that the speed of Mike's boat in still water was 20 miles/hour and the speed of the current in Lazy River was 4 miles/hour.

Extra Practice

1. Elias and Allison found 26 coins. The coins were either quarters or nickels. The total value they had was $4.70. How many quarters and nickels did they find?

2. A bulk food store wants to make a 10-pound box of snack bars that sells for $50. The box will contain large snack bars that sell for $6.00 per pound and small snack bars that sell for $4.00 per pound. How many pounds of each snack bar should they include in the box to obtain the desired mixture?

3. Ronaldo and Lee both work at the same construction site. Ronaldo worked for 5 hours and Lee worked for 6 hours. Together they hammered 875 nails. The next day, Ronaldo worked for 8 hours and Lee worked for 5 hours. Together, they hammered 1055 nails. How many nails does Ronaldo hammer per hour? How many nails does Lee hammer per hour?

4. The present population of Anytown is 45,600. The town is growing at the rate of 400 people per year. The population of the city of Anyplace is 53,100 and is increasing at the rate of 100 people per year. How many years will it be until the populations of Anytown and Anyplace are the same? What will each population be?

Concept Check

A new company is making a drink that is 45% pure fruit juice. They make a test batch of 500 gallons of the new drink. They are using some juice that is 50% pure fruit juice and some juice that is 30% pure fruit juice. They want to find out how many gallons of each of these two types they will need. Explain how you would set up a system of two equations using the variables x and y to solve this problem.

MATH COACH

Mastering the skills you need to do well on the test.

Watch the **MATH COACH** videos in MyMathLab® or on YouTube™ while you work the problems below. These helpful hints will help you avoid making common errors on test problems.

Solving a System of Equations by the Substitution Method—Problem 1 Solve by the substitution method.

$$3x - y = -5$$
$$-2x + 5y = -14$$

> **Helpful Hint:** If one equation contains $-y$, it is easier to solve for y by adding $+y$ to each side of the equation. Use this result to substitute for y in the second equation.

Did you add y to each side of the first equation and then add 5 to each side to obtain $3x + 5 = y$? Yes ____ No ____

If you answered No, consider why solving for y in the first equation is the most logical first step when using the substitution method to solve this system.

Did you substitute $(3x + 5)$ for y in the second equation to obtain $-2x + 5(3x + 5) = -14$ and then simplify and solve for x? Yes ____ No ____

If you answered No, stop and perform these steps. Remember to substitute your final value of x into one of the original equations to find y.

If you answered Problem 1 incorrectly, go back and rework the problem using these suggestions.

Solving a system of Equations with Fractional Coefficients—Problem 7

Solve by any method.
$$\frac{2}{3}x - \frac{1}{5}y = 2$$
$$\frac{4}{3}x + 4y = 4$$

> **Helpful Hint:** Find the LCD of the fractions in the top equation. Multiply each term of that equation by this LCD. Find the LCD of the fractions in the bottom equation. Multiply each term of that equation by this LCD. Now use these two new equations to solve the system.

Did you find that 15 is the LCD of the top equation? Did you multiply all three terms of the top equation by 15 to obtain $10x - 3y = 30$? Yes ____ No ____

If you answered No to theses questions, consider why the LCD is 15 and then carefully multiply each term of the top equation by 15 to find the correct result.

Did you find that 3 is the LCD of the bottom equation? Did you multiply all three terms of the bottom equation by 3 to obtain $4x + 12y = 12$? Yes ____ No ____

If you answered No to these questions, consider why the LCD is 3 and then carefully multiply each term of the bottom equation by 3 to find the correct result. Then use these two new equations to solve the system.

Now go back and rework the problem again using these suggestions.

Solving an Inconsistent or Dependent System of Equations—Problem 9

Solve by any method. If there is not one solution to a system, state why.

$$5x - 2 = y$$
$$10x = 4 + 2y$$

Helpful Hint: When you try to solve a system and get a false statement such as $3 = 0$, then the system has no solution and is inconsistent. When you try to solve a system and get a statement that is always true such as $4 = 4$, then the system has an infinite number of solutions and is dependent.

Since the top equation is already solved for y, it makes sense to use the substitution method.

Did you substitute $(5x - 2)$ for y in the bottom equation to obtain $10x = 4 + 2(5x - 2)$? Yes _____ No _____

If you answered No, go back and make this substitution.

Did you simplify the equation to obtain $0 = 0$?
Yes _____ No _____

If you used the addition method, you should still obtain $0 = 0$. Did you determine that this system has an infinite number of solutions?

Yes _____ No _____

If you answered No to these questions, examine your work for any errors and review the definition of inconsistent and dependent systems. In your final description, state whether the system is inconsistent or dependent and also state whether the system has no solution or an infinite number of solutions.

If you answered Problem 9 incorrectly, go back and rework the problem using these suggestions.

Solving a System of Equations with Decimal Coefficients—Problem 10

Solve by any method.
$$0.3x + 0.2y = 0$$
$$1.0x + 0.5y = -0.5$$

Helpful Hint: Since the decimals are tenths, multiply each term of each equation by 10 to find an equivalent system of equations without decimal coefficients. Then solve the system.

Did you multiply both equation by 10 to obtain the system
$$3x + 2y = 0$$
$$10 + 5y = -5?$$
Yes _____ No _____

If you answered No, carefully go through each equation and move the decimal point one place to the right for each number. This represents multiplying by 10. See if you get the above result.

With the new system, a good approach is to multiply the top equation by 5 and the bottom equation by −2. If you follow these steps, do you get the system
$$15x + 10y = 0$$
$$-20x - 10y = 10?$$
Yes _____ No _____

If you answered No, stop and carefully multiply every term of the top equation by 5. Remember that $5 \times 0 = 0$.

Next, multiply every term of the bottom equation by −2 and complete the problem. Watch out for + and − signs. Remember to substitute your final value for x into either of the original equations to find y.

Now go back and rework the problem using these suggestions.

144

Chapter 5 Exponents and Polynomials
5.1 The Rules of Exponents

Vocabulary

exponent • base • exponential expression • the product rule
numerical coefficient • the quotient rule

1. In the exponential expression x^a, x is called the _____.

2. In the exponential expression x^a, a is called the _____.

3. Simplifying the exponential expression $\dfrac{x^a}{x^b}$ requires using the _____.

4. A(n) _____ is a number that is multiplied by a variable, such as the 4 in $4x^2$.

Example	Student Practice
1. Multiply.	**2.** Multiply.
(a) $x^3 \cdot x^6$	**(a)** $z^4 \cdot z$
The expressions have the same base so we can add the exponents. $$x^3 \cdot x^6 = x^{3+6} = x^9$$	
(b) $x \cdot x^5$	**(b)** $2^2 \cdot 2^6$
Every variable that does not have a written exponent is understood to have an exponent of 1. Thus, $x = x^1$. $$x \cdot x^5 = x^{1+5} = x^6$$	

Vocabulary Answers: 1. base 2. exponent 3. quotient rule 4. numerical coefficient

Example	Student Practice
3. Multiply. $(5ab)\left(-\dfrac{1}{3}a\right)(9b^2)$	**4.** Multiply. $(-4x^3)(xy^2)(2x^2y)$

Multiply the numerical coefficients and group like bases.

$$(5ab)\left(-\frac{1}{3}a\right)(9b^2)$$

$$= (5)\left(-\frac{1}{3}\right)(9)(a \cdot a)(b \cdot b^2)$$

Use the rule for multiplying expression with exponents. Add the exponents.

$$(5)\left(-\frac{1}{3}\right)(9)(a \cdot a)(b \cdot b^2) = -15a^2b^3$$

Example	Student Practice
5. Divide. $\dfrac{2^{16}}{2^{11}}$	**6.** Divide. $\dfrac{x^9}{x^5}$

The expressions have the same base so we can subtract the exponents.

$$\frac{2^{16}}{2^{11}} = 2^{16-11} = 2^5$$

Example	Student Practice
7. Divide. $\dfrac{12^{17}}{12^{20}}$	**8.** Divide. $\dfrac{n^6}{n^{10}}$

The expressions have the same base so we can subtract the exponents. Notice that the larger exponent is in the denominator.

$$\frac{12^{17}}{12^{20}} = \frac{1}{12^{20-17}} = \frac{1}{12^3}$$

Example	Student Practice
9. Simplify. $\dfrac{4x^0 y^2}{8^0 y^5 z^3}$	**10.** Simplify. $\dfrac{\left(2y^4\right)\left(3x^2 y\right)}{12x^0 y^{10}}$

Any number (except 0) to the 0 power equals 1.

$$\dfrac{4x^0 y^2}{8^0 y^5 z^3} = \dfrac{4(1) y^2}{(1) y^5 z^3}$$

$$= \dfrac{4y^2}{y^5 z^3}$$

$$= \dfrac{4}{y^3 z^3}$$

11. Simplify.	**12.** Simplify.
(a) $\left(x^3\right)^5$	**(a)** $\left(y^8\right)^4$

We are raising a power to a power, so multiply exponents.

$$\left(x^3\right)^5 = x^{3 \cdot 5} = x^{15}$$

(b) $(-1)^8$

(b) $(-1)^{13}$

Since n is even, a positive number results.

$$\left(-1^8\right) = +1$$

13. Simplify. $\left(\dfrac{x}{y}\right)^5$	**14.** Simplify. $\left(\dfrac{y}{z^2}\right)^3$

The fraction within the parentheses is raised to a power. Raise both the numerator and the denominator to that power.

$$\left(\dfrac{x}{y}\right)^5 = \dfrac{x^5}{y^5}$$

Example	Student Practice
15. Simplify. $\left(\dfrac{-3x^2z^0}{y^3}\right)^4$	**16.** Simplify. $\left(\dfrac{4x^5y}{-16x^0y^3}\right)^3$

Simplify inside the parentheses first. Apply the rules for raising a power to a power and simplify.

$$\left(\frac{-3x^2z^0}{y^3}\right)^4 = \left(\frac{-3x^2}{y^3}\right)^4$$

$$= \frac{(-3)^4 x^8}{y^{12}} = \frac{81x^8}{y^{12}}$$

Extra Practice

1. Multiply. Leave your answer in exponent form. $\left(12x^4y^2\right)\left(2x^2y^3\right)$

2. Simplify. Leave your answer in exponent form. Assume that all variables in any denominator are nonzero. $\dfrac{9a^3}{a^3}$

3. Simplify. Leave your answer in exponent form. Assume that all variables in any denominator are nonzero. $\left(\dfrac{b}{b^3}\right)^2$

4. Simplify. Leave your answer in exponent form. Assume that all variables in any denominator are nonzero. $\left(\dfrac{9x^2y^0}{14x^5}\right)^2$

Concept Check

Explain the steps you would need to follow to simplify the expression. $\dfrac{\left(4x^3\right)^2}{\left(2x^4\right)^3}$

Chapter 5 Exponents and Polynomials
5.2 Negative Exponents and Scientific Notation

Vocabulary
exponent • base • negative exponent • scientific notation • significant digits

1. All digits in a number, excluding the zeros that allow the decimal point to be properly located, are considered _____.

2. In the number 1.23×10^5, "10" is referred to as the _____.

3. A useful way to express very large or very small numbers is to use _____.

4. When using scientific notation to express a number less than zero, it is necessary to use a _____ on the base of ten.

Example	**Student Practice**
1. Evaluate. 3^{-2}	**2.** Evaluate. 7^{-4}
First write the expression with a positive exponent. Then evaluate.	
$3^{-2} = \dfrac{1}{3^2} = \dfrac{1}{9}$	
3. Simplify. Write the expression with no negative exponents. $\left(3x^{-4}y^2\right)^{-3}$	**4.** Simplify. Write the expression with no negative exponents. $\left(2xy^{-2}z^3\right)^{-4}$
Use the power to power rule.	
$\left(3x^{-4}y^2\right)^{-3} = 3^{-3}x^{12}y^{-6}$	
Now write the expression with positive exponents only and simplify.	
$3^{-3}x^{12}y^{-6} = \dfrac{x^{12}}{3^3 y^6} = \dfrac{x^{12}}{27y^6}$	

Vocabulary Answers: 1. significant digits 2. base 3. scientific notation 4. negative exponent

Example	Student Practice
5. Write in scientific notation. $157,000,000$	**6.** Write in scientific notation. 2564

5. Write in scientific notation. $157,000,000$

Move the decimal point 8 places to the left and multiply by $100,000,000$.

$$157,000,000 = 1.\underbrace{57000000}_{\text{8 places}} \times 1\underbrace{00000000}_{\text{8 zeros}}$$

$$= 1.57 \times 10^8$$

6. Write in scientific notation. 2564

---

7. Write in scientific notation.

(a) 0.061

Move the decimal point 2 places to the right and multiply by 10^{-2}.

$0.061 = 6.1 \times 10^{-2}$

(b) 0.000052

$0.000052 = 5.2 \times 10^{-5}$

8. Write in scientific notation.

(a) 0.0013

(b) 0.000001

---

9. Write in decimal notation.

(a) 1.568×10^2

The exponent 2 tells us to move the decimal point 2 places to the right.

$1.568 \times 10^2 = 156.8$

(b) 7.432×10^{-3}

The exponent -3 tells us to move the decimal point 3 places to the left.

$$7.432 \times 10^{-3} = 7.432 \times \frac{1}{1000}$$

$$= 0.007432$$

10. Write in decimal notation.

(a) 5.28×10^{-4}

(b) 3.2214×10^6

150

Example	Student Practice

11. The approximate distance from Earth to the star Polaris is 208 parsecs (pc). A parsec is a distance of approximately 3.09×10^{13} km. How long would it take a space probe traveling at 40,000 km/hr to reach the star? Round to three significant digits.

Understand the problem. We need to change the distance from parsecs to kilometers using the given relationship.

$$208 \text{ pc} = \frac{(208 \text{ pc})(3.09 \times 10^{13} \text{ km})}{1 \text{ pc}}$$

$$= 642.72 \times 10^{13}$$

Write an equation using the distance formula, $\text{distance} = \text{rate} \times \text{time}$, or $d = r \times t$.

Substitute known values and change all given values to scientific notation.

$$6.4272 \times 10^{15} \text{ km} = \frac{4 \times 10^4 \text{ km}}{1 \text{ hr}} \times t$$

Multiply both sides by the reciprocal of $\dfrac{4 \times 10^4 \text{ km}}{1 \text{ hr}}$ and simplify.

$$6.4272 \times 10^{15} \text{ km} \times \frac{1 \text{ hr}}{4 \times 10^4 \text{ km}} = t$$

$$\frac{\left(6.4272 \times 10^{15} \, \cancel{\text{km}}\right)(1 \text{ hr})}{4 \times 10^4 \, \cancel{\text{km}}} = t$$

$$1.6068 \times 10^{11} \text{ hr} = t$$

Round to three significant figures.
$1.6068 \times 10^{11} \text{ hr} \approx 1.61 \times 10^{11} \text{ hr}$

The check is left to the student.

12. Use the information in example **15** to answer the following. How long would it take the space probe to reach a star that is 500 parsecs from Earth? Round to three significant digits.

Extra Practice

1. Simplify. Express your answer with positive exponents. Assume that all variables are nonzero. 7^{-3}

2. Simplify. Express your answer with positive exponents. Assume that all variables are nonzero. $\left(\dfrac{3a^3b^{-2}}{c^3}\right)^{-4}$

3. Write in decimal notation. 6.34×10^3

4. Evaluate by using scientific notation and the laws of exponents. Leave your answer in scientific notation. $\dfrac{0.0046}{0.023}$

Concept Check

Explain how you would simplify a problem like the following so that your answer has only positive exponents. $\left(4x^{-3}y^4\right)^{-3}$

Chapter 5 Exponents and Polynomials
5.3 Fundamental Polynomial Operations

Vocabulary

polynomial • multivariable polynomial • degree of a term • degree of a polynomial
monomial • binomial • trinomial • decreasing order • evaluate

1. A(n) _____ is a polynomial with two terms.

2. In a term, the sum of the exponents of all of the variables is called the _____.

3. When a polynomial is written in _____, the value of each exponent decreases as we move from left to right.

4. The highest degree of all of the terms in a polynomial is called the _____.

Example	**Student Practice**
1. State the degree of the polynomial, and whether it is a monomial, a binomial, or a trinomial.	**2.** State the degree of the polynomial, and whether it is a monomial, a binomial, or a trinomial.
(a) $5xy + 3x^3$	**(a)** $7y^3z^4$
This polynomial is of degree 3. It has two terms, so it is a binomial.	
(b) $-7a^5b^2$	**(b)** $x^2 + 5x - 4$
The sum of the exponents is $5 + 2 = 7$. Therefore, this polynomial is of degree 7. It has one term, so it is a monomial.	
(c) $8x^4 - 9x - 15$	**(c)** $2xy^6 + 7x^2$
This polynomial is of degree 4. It has three terms, so it is a trinomial.	

Vocabulary Answers: 1. binomial 2. degree of a term 3. decreasing order 4. degree of a polynomial

Example	Student Practice
3. Add. $\left(5x^2 - 6x - 12\right) + \left(-3x^2 - 9x + 5\right)$	**4.** Add. $\left(-x^2 + x - 5\right) + \left(5x^2 - 9x + 10\right)$

3. Add. $\left(5x^2 - 6x - 12\right) + \left(-3x^2 - 9x + 5\right)$

Group like terms.

$$\left(5x^2 - 6x - 12\right) + \left(-3x^2 - 9x + 5\right)$$
$$= \left[5x^2 + \left(-3x^2\right)\right] + \left[-6x + \left(-9x\right)\right] + \left[-12 + 5\right]$$

Add like terms.

$$\left[(5-3)x^2\right] + \left[(-6-9)x\right] + \left[-12 + 5\right]$$
$$= 2x^2 + \left(-15x\right) + \left(-7\right)$$
$$= 2x^2 - 15x - 7$$

---

5. Add. $\left(\dfrac{1}{2}x^2 - 6x + \dfrac{1}{3}\right) + \left(\dfrac{1}{5}x^2 - 2x - \dfrac{1}{2}\right)$

6. Add. $\left(\dfrac{3}{4}x^2 - 2x + \dfrac{1}{8}\right) + \left(x^2 + \dfrac{2}{5}x + \dfrac{1}{2}\right)$

The numerical coefficients of polynomials may be any real number. Thus, polynomials may have numerical coefficients that are decimals or fractions.

To add, first group like terms.

$$\left(\dfrac{1}{2}x^2 - 6x + \dfrac{1}{3}\right) + \left(\dfrac{1}{5}x^2 - 2x - \dfrac{1}{2}\right)$$
$$= \left[\dfrac{1}{2}x^2 + \dfrac{1}{5}x^2\right] + \left[-6x + \left(-2x\right)\right] + \left[\dfrac{1}{3} + \left(-\dfrac{1}{2}\right)\right]$$

Add like terms.

$$\left[\left(\dfrac{1}{2} + \dfrac{1}{5}\right)x^2\right] + \left[(-6-2)x\right] + \left[\dfrac{1}{3} + \left(-\dfrac{1}{2}\right)\right]$$
$$= \left[\left(\dfrac{5}{10} + \dfrac{2}{10}\right)x^2\right] + \left(-8x\right) + \left[\dfrac{2}{6} - \dfrac{3}{6}\right]$$
$$= \dfrac{7}{10}x^2 - 8x - \dfrac{1}{6}$$

Example	Student Practice
7. Subtract. $\left(7x^2 - 6x + 3\right) - \left(5x^2 - 8x - 12\right)$	**8.** Subtract. $\left(-5x^2 - 3x + 1\right) - \left(-2x^2 + 7x - 4\right)$

We change the sign of each term in the second polynomial and then add.

$$\left(7x^2 - 6x + 3\right) - \left(5x^2 - 8x - 12\right)$$

$$= \left(7x^2 - 6x + 3\right) + \left(-5x^2 + 8x + 12\right)$$

$$= \left(7 - 5\right)x^2 + \left(-6 + 8\right)x + \left(3 + 12\right)$$

$$= 2x^2 + 2x + 15$$

9. Automobiles sold in the United States have become more fuel efficient over the years due to regulations from Congress. The number of miles per gallon obtained by the average automobile in the United States can be described by the polynomial $0.3x + 12.9$, where x is the number of years since 1970. (*Source:* U.S. Federal Highway Administration.) Use this polynomial to estimate the number of miles per gallon obtained by the average automobile in 1972.

The year 1972 is two years later than 1970, so $x = 2$.

Thus, the number of miles per gallon obtained by the average automobile in 1972 can be estimated by evaluating $0.3x + 12.9$ when $x = 2$.

$$0.3(2) + 12.9 = 0.6 + 12.9$$

$$= 13.5$$

We estimate that the average car in 1972 obtained 13.5 miles per gallon.

10. Use the information in example **9** to answer the following. Estimate the number of miles per gallon that will be obtained by the average automobile in the United States in 2015.

Extra Practice

1. State the degree of the polynomial and whether it is a monomial, a binomial, or a trinomial. $23x^6 - 14x^3 + 5$

2. Add. $(6.4x - 3) + (4.4x - 11)$

3. Subtract.

   $$(2r^4 - 3r^2 + 14) - (-3r^4 - 2r^2 + 6)$$

4. Buses sold in the U.S. have become more fuel efficient over the years. The number of miles per gallon obtained by a certain type of bus can be described by the polynomial $0.37x + 5.31$, where x is the number of years since 1970. Use this polynomial to estimate the number of miles per gallon obtained by this type of bus in 1980.

Concept Check

Explain how you would determine the degree of the following polynomial and how you would decide if it is a monomial, a binomial, or a trinomial. $2xy^2 - 5x^3y^4$

Chapter 5 Exponents and Polynomials
5.4 Multiplying Polynomials

Vocabulary

polynomial • monomial • binomial • distributive property • FOIL

1. A _____ is a polynomial with only one term.

2. The process of using the distributive property to multiply two binomials is often referred
   to as _____.

3. The _____ states that for all real numbers a, b, and c, $a(b+c) = ab + ac$.

Example	Student Practice
1. Multiply. $3x^2(5x-2)$	**2.** Multiply. $5y^2(-y+4)$

1. Multiply. $3x^2(5x-2)$

Use the distributive property and
multiply each term by $3x^2$.

$$3x^2(5x-2) = 3x^2(5x) + 3x^2(-2)$$

Simplify.

$$3x^2(5x) + 3x^2(-2)$$
$$= (3\cdot5)(x^2 \cdot x) + (3)(-2)x^2$$
$$= 15x^3 - 6x^2$$

2. Multiply. $5y^2(-y+4)$

---

3. Multiply. $(x^2 - 2x + 6)(-2xy)$

Use the distributive property and
multiply each term by $-2xy$.

$$(x^2 - 2x + 6)(-2xy)$$
$$= -2x^3y + 4x^2y - 12xy$$

4. Multiply. $(x^2 - 7x - 10)(2x^2y)$

Vocabulary Answers: 1. monomial 2. FOIL 3. distributive property

Example	Student Practice
5. Multiply. $(2x-1)(3x+2)$	**6.** Multiply. $(3x-1)(5x+3)$

Multiply the First terms, $2x$ and $3x$.

$$(2x)(3x)=6x^2$$

Multiply the Outer terms, $2x$ and 2.

$$(2x)(2)=4x$$

Multiply the Inner terms, -1 and $3x$.

$$(-1)(3x)=-3x$$

Multiply the Last terms, -1 and 2.

$$(-1)(2)=-2$$

Add the results and combine like terms.
  First + Outer + Inner + Last

$$=6x^2 + 4x - 3x - 2$$

$$=6x^2+x-2$$

7. Multiply. $(3x+2y)(5x-3z)$	**8.** Multiply. $(7x-3y)(x+2z)$

Multiply the First terms, $3x$ and $5x$.

$$(3x)(5x)=15x^2$$

Multiply the Outer terms, $3x$ and $-3z$.

$$(3x)(-3z)=-9xz$$

Multiply the Inner terms, $2y$ and $5x$.

$$(2y)(5x)=10xy$$

Multiply the Last terms, $2y$ and $-3z$.

$$(2y)(-3z)=-6yz$$

Add the results and combine like terms.
$$(3x+2y)(5x-3z)$$

$$=15x^2-9xz+10xy-6yz$$

Example	Student Practice
9. Multiply. $(7x-2y)^2$	**10.** Multiply. $(4x+3y)^2$

When we square a binomial, it is the same as multiplying the binomial by itself.

$$(7x-2y)(7x-2y)$$

Multiply the First terms, $7x$ and $7x$.
$$(7x)(7x)=49x^2$$

Multiply the Outer terms, $7x$ and $-2y$.
$$(7x)(-2y)=-14xy$$

Multiply the Inner terms, $-2y$ and $7x$.
$$(-2y)(7x)=-14xy$$

Multiply the Last terms, $-2y$ and $-2y$.
$$(-2y)(-2y)=4y^2$$

Add the results and combine like terms.

$$(7x-2y)^2 = 49x^2 -14xy -14xy +4y^2$$
$$= 49x^2 - 28xy + 4y^2$$

11. Multiply. $\left(3x^2+4y^3\right)\left(2x^2+5y^3\right)$	**12.** Multiply. $\left(10x^2+5y^4\right)\left(2x^2-3y^4\right)$

Use the FOIL method and the rules for multiplying expressions with exponents.

$$\left(3x^2+4y^3\right)\left(2x^2+5y^3\right)$$
$$= 6x^4 +15x^2y^3 +8x^2y^3 +20y^6$$
$$= 6x^4 +23x^2y^3 +20y^6$$

Example	Student Practice
13. The width of a living room is $(x+4)$ feet. The length of the room is $(3x+5)$ feet. What is the area of the room in square feet?	**14.** The width of a brick patio is $(2x+1)$ feet. The length of the patio is $(4x-2)$ feet. What is the area of the patio in square feet?

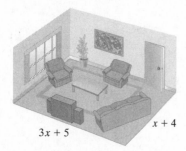

$3x + 5$ $x + 4$

Use the area formula and solve.

$$A = (\text{length})(\text{width})$$
$$A = (3x+5)(x+4)$$
$$A = 3x^2 +12x+5x+20$$
$$A = 3x^2 +17x+20$$

There are $\left(3x^2 +17x+20\right)$ square feet in the room.

Extra Practice

1. Multiply. $5x\left(-2x^3 +3x\right)$

2. Multiply. $(x-3)(x-11)$

3. Multiply. $(3x-7y)(5x+8y)$

4. Multiply. $\left(2x^2 -3y^3\right)\left(4x^2 +5y^3\right)$

Concept Check

Explain how you would multiply $(7x-3)^2$.

Chapter 5 Exponents and Polynomials
5.5 Multiplication: Special Cases

Vocabulary
polynomial   •   term   •   binomial   •   FOIL   •   square of a sum
square of a difference   •   vertical multiplication   •   horizontal multiplication

1. One method of multiplying polynomials that uses a method similar to that used in arithmetic for multiplying whole numbers is called _____.

2. A binomial of the form $(a-b)^2$ is referred to as the _____.

3. One way to multiply polynomials with more than two terms is to use _____ horizontal multiplication.

Example	**Student Practice**
1. Multiply. $(7x+2)(7x-2)$ Use the rule for multiplying a sum and a difference. $(a+b)(a-b)=a^2-b^2$ Here, $a=7x$ and $b=2$. $(7x+2)(7x-2)=(7x)^2-(2)^2$ $\qquad\qquad\qquad =49x^2-4$ The check is left to the student.	**2.** Multiply. $(8y+4)(8y-4)$
3. Multiply. $(5x-8y)(5x+8y)$ Use the rule for multiplying a sum and a difference. Here, $a=5x$ and $b=8y$. $(5x-8y)(5x+8y)=(5x)^2-(8y)^2$ $\qquad\qquad\qquad\qquad =25x^2-64y^2$	**4.** Multiply. $(2x-9y)(2x+9y)$

Vocabulary Answers: 1. vertical multiplication 2. square of a difference 3. FOIL

Example	Student Practice
5. Multiply. $(5y-2)^2$ Use a rule for a binomial squared. $(a+b)^2 = a^2 + 2ab + b^2$ $(a-b)^2 = a^2 - 2ab + b^2$ Here, $a = 5y$ and $b = 2$. $(5y-2)^2$ $= (5y)^2 - (2)(5y)(2) + (2)^2$ $= 25y^2 - 20y + 4$	**6.** Multiply. $(7x+y)^2$
7. Multiply vertically. $(3x^3 + 2x^2 + x)(x^2 - 2x - 4)$ Place one polynomial over the other. Find the partial products and line them up under the original polynomials. Note that the answers for each partial product are placed so that like terms are underneath each other. $$\begin{array}{r} 3x^3 + 2x^2 + x \\ x^2 - 2x - 4 \\ \hline -12x^3 - 8x^2 - 4x \\ -6x^4 - 4x^3 - 2x^2 \\ 3x^5 - 2x^4 - x^3 \end{array}$$ Find the sum of the three partial products. $$\begin{array}{r} 3x^3 + 2x^2 + x \\ x^2 - 2x - 4 \\ \hline -12x^3 - 8x^2 - 4x \\ -6x^4 - 4x^3 - 2x^2 \\ 3x^5 - 2x^4 - x^3 \\ \hline 3x^5 - 4x^4 - 15x^3 - 10x^2 - 4x \end{array}$$	**8.** Multiply vertically. $(x^2 + 3x - 6)(2x^2 - 4x - 5)$

Example	Student Practice
9. Multiply horizontally. $$\left(x^2+3x+5\right)\left(x^2-2x-6\right)$$	**10.** Multiply horizontally. $$\left(x^2-8x+10\right)\left(x^2+x+4\right)$$

Use the distributive property repeatedly.

$$\left(x^2+3x+5\right)\left(x^2-2x-6\right)$$

$$=x^2\left(x^2-2x-6\right)+3x\left(x^2-2x-6\right)$$

$$+5\left(x^2-2x-6\right)$$

$$=x^4-2x^3-6x^2+3x^3-6x^2-18x$$

$$+5x^2-10x-30$$

$$=x^4+x^3-7x^2-28x-30$$

---

11. Multiply. $(2x-3)(x+2)(x+1)$

12. Multiply. $(x-2)(x+3)(2x+4)$

Multiply the first pair of binomials. Note that it does not matter which two binomials are multiplied first.

$$(2x-3)(x+2)=2x^2+4x-3x-6$$

$$=2x^2+x-6$$

Replace the first two factors with their resulting product.

$$\left(2x^2+x-6\right)(x+1)$$

Multiply again and combine like terms.

$$\left(2x^2+x-6\right)(x+1)$$

$$=\left(2x^2+x-6\right)x+\left(2x^2+x-6\right)1$$

$$=2x^3+x^2-6x+2x^2+x-6$$

$$=2x^3+3x^2-5x-6$$

Extra Practice

1. Multiply. Use the special formula that applies. $(x+10)(x-10)$

2. Multiply. Use the special formula that applies. $(5a+3b)(5a-3b)$

3. Multiply. $(8x^2-2x+3)(3x+1)$

4. Multiply. $(x+2)(x-1)(x-5)$

Concept Check

Using the formula $(a+b)^2 = a^2 + 2ab + b^2$, explain how to multiply $(6x-9y)^2$.

Chapter 5 Exponents and Polynomials
5.6 Dividing Polynomials

Vocabulary
polynomial • subtraction • long division • descending order • binomial • monomial

1. To divide a polynomial by a _____, divide each term of the numerator by the denominator, then write the sum of the results.

2. _____ is a process used to divide polynomials when the divisor has two or more terms.

3. When dividing a polynomial by a binomial, you must first place the terms of each in _____, inserting a 0 for any missing terms.

4. When performing long division with polynomials, take great care on the_____ step when negative numbers are involved.

Example	Student Practice
1. Divide. $\dfrac{8y^6 - 8y^4 + 24y^2}{8y^2}$	**2.** Divide. $\dfrac{21x^5 + 7x^3 - 28x^2}{7x^2}$

Divide each term of the polynomial by the monomial.

$$\frac{8y^6 - 8y^4 + 24y^2}{8y^2} = \frac{8y^6}{8y^2} - \frac{8y^4}{8y^2} + \frac{24y^2}{8y^2}$$

Use the property $\dfrac{x^a}{x^b} = x^{a-b}$ to divide each term.

$$\frac{8y^6}{8y^2} - \frac{8y^4}{8y^2} + \frac{24y^2}{8y^2} = y^4 - y^2 + 3$$

Vocabulary Answers: 1. monomial 2. descending order 3. long division 4. subtraction

Example	Student Practice
3. Divide. $\left(x^3 + 5x^2 + 11x + 4\right) \div (x + 2)$	**4.** Divide. $\left(2x^3 + 4x^2 + 3x + 6\right) \div (x + 1)$

Notice the terms are already in descending order with no missing terms. Divide the first term of the polynomial by the first term of the binomial.

$$x + 2 \overline{)\begin{array}{l} x^2 \\ x^3 + 5x^2 + 11x + 4 \end{array}}$$

Multiply x^2 by $x + 2$ and subtract the result from the first two terms.

$$x + 2 \overline{)\begin{array}{l} x^2 + 3x \\ x^3 + 5x^2 + 11x + 4 \\ \underline{x^3 + 2x^2} \\ 3x^2 + 11x \end{array}}$$

Continue this process until the degree of the remainder is less than the degree of the divisor.

$$x + 2 \overline{)\begin{array}{l} x^2 + 3x + 5 \\ x^3 + 5x^2 + 11x + 4 \\ \underline{x^3 + 2x^2} \\ 3x^2 + 11x \\ \underline{3x^2 + 6x} \\ 5x + 4 \\ \underline{5x + 10} \\ -6 \end{array}}$$

The remainder is written as the numerator of a fraction that has the binomial divisor as its denominator.

$$x^2 + 3x + 5 + \frac{-6}{x + 2}$$

The check is left to the student.

Example	Student Practice
5. Divide. $\left(5x^3 - 24x^2 + 9\right) \div (5x+1)$	**6.** Divide. $\left(8x^3 - 8x + 5\right) \div (2x+1)$

Insert $0x$ into the polynomial to represent the missing x-term.

$$\left(5x^3 - 24x^2 + 0x + 9\right) \div (5x+1)$$

Divide the first term of the polynomial, $5x^3$, by the first term of the binomial, $5x$.

$$
\begin{array}{r}
x^2 \\
5x+1{\overline{\smash{\big)}\,5x^3 - 24x^2 + 0x + 9}} \\
\underline{5x^3 + x^2} \\
-25x^2
\end{array}
$$

Divide $-25x^2$ by $5x$ and then divide $5x$ by $5x$. Be cautious with the negative signs.

$$
\begin{array}{r}
x^2 - 5x + 1 \\
5x+1{\overline{\smash{\big)}\,5x^3 - 24x^2 + 0x + 9}} \\
\underline{5x^3 + x^2} \\
-25x^2 + 0x \\
\underline{-25x^2 - 5x} \\
5x + 9 \\
\underline{5x + 1} \\
8
\end{array}
$$

Write the remainder as the numerator of a fraction that has the binomial divisor as its denominator.

The answer is $x^2 - 5x + 1 + \dfrac{8}{5x+1}$.

The check is left to the student.

Example	Student Practice
7. Divide and check.	**8.** Divide and check.
$\left(12x^3 - 11x^2 + 8x - 4\right) \div \left(3x - 2\right)$	$\left(8x^3 - 4x^2 + 6\right) \div \left(2x - 3\right)$

$$\begin{array}{r} 4x^2 - x + 2 \\ 3x-2 \overline{)12x^3 - 11x^2 + 8x - 4} \\ \underline{12x^3 - 8x^2} \\ -3x^2 + 8x \\ \underline{-3x^2 + 2x} \\ 6x - 4 \\ \underline{6x - 4} \\ 0 \end{array}$$

Check the answer.

$$\left(3x - 2\right)\left(4x^2 - x + 2\right) = 12x^3 - 11x^2 + 8x - 4$$

Extra Practice

1. Divide. $\dfrac{12a^7 - 4a^5 + 8a^3 - 2a^2}{2a^2}$

2. Divide. $\left(12y^4 - 18y^3 + 27y^2\right) \div 3y^2$

3. Divide and check. $\dfrac{3x^3 - 5x^2 + 7x - 5}{x - 1}$

Divide and check. $\dfrac{y^3 - 2y - 4}{y - 1}$

Concept Check

Explain how you would check if $x^2 + 2x + 8 + \dfrac{13}{x - 2}$ is the correct answer to the problem $\left(x^3 + 4x - 3\right) \div \left(x - 2\right)$. Perform the check. Does the answer check?

MATH COACH

Mastering the skills you need to do well on the test.

Watch the **MATH COACH** videos in MyMathLab® or on You Tube™ while you work the problems below. These helpful hints will help you avoid making common errors on test problems.

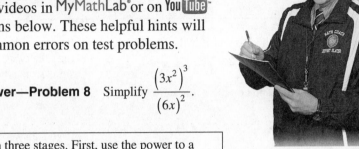

Raising Monomials to a Power—Problem 8   Simplify $\dfrac{\left(3x^2\right)^3}{\left(6x\right)^2}$.

Helpful Hint: Do the problem in three stages. First, use the power to a power rule to raise the numerator to the third power. Second, raise the denominator to the second power. Then divide the monomials using the rules of exponents. Be sure to simplify any fractions.

Did you use the power to a power rule to raise both 3^1 and x^2 to the third power in the numerator and both 6^1 and x^1 to the second power in the denominator?
Yes ____ No ____

If you answered No, stop and review the power to a power rule before completing these steps again.

Did you remember to simplify the fraction $\dfrac{27}{36}$?

Yes ____ No ____

Finally, did you remember to use the quotient rule to subtract the exponents in the x terms?
Yes ____ No ____

If you answered No to either of these questions, go back and examine your work carefully before completing these steps again.

If you answered Problem 8 incorrectly, go back and rework the problem using these suggestions.

Simplifying Monomials Involving Negative Exponents—Problem 11

Simplify and write with only positive exponents. $\dfrac{3x^{-3}y^2}{x^{-4}y^{-5}}$

Helpful Hint: First, use the definition of a negative exponent to rewrite the expression using only positive exponents. Then use the rules for exponents to simplify the resulting expression.

Did you remove the negative exponents by rewriting the expression as $\dfrac{3x^4y^2y^5}{x^3}$? Yes ____ No ____

If you answered No, review the definition of negative exponents in Section 4.2 and complete this step again.

Did you use the quotient rule to simplify the x terms and the product rule to simplify the y terms? Yes ____ No ____

If you answered No, review the rules for exponents in Sections 4.1 and 4.2 and simplify the expression again.

Now go back and rework the problem using these suggestions.

Multiplying Three Binomials—Problem 20 Multiply $(3x+2)(2x+1)(x-3)$.

Helpful Hint: A good approach is to start by multiplying the first two binomials. Then multiply that result by the third binomial. Be careful to avoid sign errors when multiplying, and be careful to write down the correct exponent for each term.

Did you use the FOIL method to multiply the first two binomials and obtain $6x^2+7x+2$? Yes _____ No _____

If you answered No, stop and complete this step.

Did you multiply the result above by $(x-3)$?
Yes _____ No _____

Did you multiply each term of $6x^2+7x+2$ by x?
Yes _____ No _____

Did you multiply each term of $6x^2+7x+2$ by -3?
Yes _____ No _____

If you answered No to any of these questions, go back and examine each step of the multiplication carefully.

Be sure to write the correct exponent each time that you multiply. Be careful to avoid sign errors when multiplying by -3. Then combine like terms before writing your final answer.

If you answered Problem 20 incorrectly, go back and rework the problem using these suggestions.

Dividing a Polynomial by a Binomial—Problem 27 Divide $(2x^3-6x-36) \div (x-3)$.

Helpful Hint: Review the procedure for dividing a polynomial by a binomial in Section 4.6. Make sure you understand each step. Be sure you understand where the expression $0x^2$ came from in the dividend. Be careful with subtraction. Write out the subtraction steps to avoid sign errors.

Did you write the division problem in the form
$x-3\overline{)2x^3+0x^2-6x-36}$? Yes _____ No _____
If you answered No, remember that every power must be represented. We must use $0x^2$ as a placeholder so that we can perform our division.

When you carried out the first step of division, did you obtain $2x^2$ as the first part of your answer?
Yes _____ No _____

When you multiplied $x-3$ by $2x^2$, and then subtracted, did you get the result $6x^2$? Yes _____ No _____

If you answered No to these questions, stop and examine your first division step carefully. Make sure that you subtracted carefully too. Remember to write out the subtraction steps: $0x^2-\left(-6x^2\right)=6x^2$.

Next, did you bring down $-6x$ from the dividend to obtain $6x^2-6x$?
Yes _____ No _____

If you answered No, go back and look at the dividend again and see how to obtain this result.

Now go back and rework the problem using these suggestions.

Chapter 6 Factoring
6.1 Removing a Common Factor

Vocabulary
factor   •   to factor   •   common factor   •   greatest common factor

1. When two or more numbers, variables, or algebraic expressions are multiplied, each is called a _____.

2. A _____ is a factor that both terms have in common.

3. When you are asked _____ a number or an algebraic expression, you are being asked, "What factors, when multiplied, will give that number or expression?"

4. When we factor, we begin by looking for the _____.

Example	Student Practice
1. Factor. **(a)** $3x - 6y$ Begin by looking for a common factor, a factor that both terms have in common. Then rewrite the expression as a product. $3x - 6y = 3(x - 2y)$ This is true because $3(x - 2y) = 3x - 6y$. **(b)** $9x + 2xy$ $9x + 2xy = x(9 + 2y)$ This is true because $x(9 + 2y) = 9x + 2xy$. Notice that factoring is using the distributive property in reverse.	**2.** Factor. **(a)** $5y + 15z$ **(b)** $12y - 5yz$

Vocabulary Answers: 1. factor 2. common factor 3. to factor 4. greatest common factor

Example	Student Practice
3. Factor $24xy + 12x^2 + 36x^3$. Remember to remove the greatest common factor.	**4.** Factor $44y^3 + 55y^2 - 11xy$. Remember to remove the greatest common factor.

3. (continued)

We start by finding the greatest common factor of 24, 12, and 36.

You may want to factor each number, or you may notice that 12 is a common factor. 12 is the greatest numerical common factor.

Notice also that x is a factor of each term. Thus, $12x$ is the greatest common factor.

$$24xy + 12x^2 + 36x^3 = 12x\left(2y + x + 3x^2\right)$$

---

5. Factor.

(a) $12x^2 - 18y^2$

Note that the largest integer that is common to both terms is 6 (not 3 or 2).

$$12x^2 - 18y^2 = 6\left(2x^2 - 3y^2\right)$$

(b) $x^2y^2 + 3xy^2 + y^3$

Although y is common to all of the terms, we factor out y^2 since 2 is the largest exponent of y that is common to all terms.

We do not factor out x, since x is not common to all of the terms.

$$x^2y^2 + 3xy^2 + y^3 = y^2\left(x^2 + 3x + y\right)$$

6. Factor.

(a) $21m^2 - 28n^2$

(b) $m^3n^2 + 9m^2n^2 + 3m^4$

172

Copyright © 2013 Pearson Education, Inc.

Example	Student Practice
7. Factor. $8x^3y + 16x^2y^2 - 24x^3y^3$	**8.** Factor. $27xy^3 - 36x^2y^2 - 9x^3y^3$

7. Factor. $8x^3y + 16x^2y^2 - 24x^3y^3$

We see that 8 is the largest integer that will divide evenly into the three numerical coefficients. We can factor x^2 out of each term. We can also factor y out of each term.

$8x^3y + 16x^2y^2 - 24x^3y^3$
$= 8x^2y\left(x + 2y - 3xy^2\right)$

Check.

$8x^2y\left(x + 2y - 3xy^2\right)$
$= 8x^3y + 16x^2y^2 - 24x^3y^3$

---

9. Factor. $9a^3b^2 + 9a^2b^2$

10. Factor. $11m^4n^2 + 11m^3n^2$

We observe that both terms contain a common factor of 9. We can also factor a^2 and b^2 out of each term. Thus, the greatest common factor is $9a^2b^2$.

$9a^3b^2 + 9a^2b^2 = 9a^2b^2\left(a + 1\right)$

---

11. Factor. $7x^2\left(2x - 3y\right) - \left(2x - 3y\right)$

12. Factor. $5y^2\left(3x + 4y\right) - \left(3x + 4y\right)$

The common factor of the terms is $\left(2x - 3y\right)$. What happens when we factor out $\left(2x - 3y\right)$? What are we left with in the second term?
Recall that $\left(2x - 3y\right) = 1\left(2x - 3y\right)$.

$7x^2\left(2x - 3y\right) - \left(2x - 3y\right)$
$= 7x^2\left(2x - 3y\right) - \mathbf{1}\left(2x - 3y\right)$
$= \left(2x - 3y\right)\left(7x^2 - 1\right)$

Example	Student Practice
13. A computer programmer is writing a program to find the total area of 4 circles. She uses the formula $A = \pi r^2$. The radii of the circles are a, b, c, and d, respectively. She wants the final answer to be in factored form with the value of π occurring only once, in order to minimize the rounding error. Write the total area of the 4 circles with a formula that has π occurring only once.	**14.** Find the total area of 3 circles using the formula $A = \pi r^2$. The radii of the circles are m, $2n$, and $3z$, respectively. Write the total area of the 3 circles with a formula that has π occurring only once.

For each circle, $A = \pi r^2$, where $r = a$, b, c, or d. We add the area of each of the 4 circles.

The total area is $\pi a^2 + \pi b^2 + \pi c^2 + \pi d^2$.

In factored form the total area =
$\pi\left(a^2 + b^2 + c^2 + d^2\right)$.

Extra Practice

1. Remove the largest possible common factor. Check your answer by multiplication. $3x^3 + 12x^2 - 21x$

2. Remove the largest possible common factor. Check your answer by multiplication. $60x^2 y + 18xy - 24x$

3. Remove the largest possible common factor. Check your answer by multiplication. $8a(x+3y) - b(x+3y)$

4. Remove the largest possible common factor. Check your answer by multiplication. $5a(4x-3) - (4x-3)$

Concept Check

Explain how you would remove the greatest common factor from the following polynomial. $36a^3 b^2 - 72a^2 b^3$

Chapter 6 Factoring
6.2 Factoring by Grouping

Vocabulary

factoring by grouping • common factor • commutative property • FOIL

1. Sometimes you will need to factor out a negative _____ from the second two terms to obtain two terms that contain the same parenthetical expression.

2. A procedure used to factor a four-term polynomial is called _____.

3. Rearrange the order using the _____ of addition.

4. To check, we multiply the two binomials using the _____ procedure.

Example	Student Practice
1. Factor. $x(x-3)+2(x-3)$	**2.** Factor. $z(2z+5)-4(2z+5)$

Observe each term: $\underbrace{x(x-3)}_{\substack{\text{first} \\ \text{term}}}+\underbrace{2(x-3)}_{\substack{\text{second} \\ \text{term}}}$

The common factor of the first and second terms is the quantity $(x-3)$.

$x(x-3)+2(x-3)=(x-3)(x+2)$

3. Factor. $2x^2+3x+6x+9$	**4.** Factor. $12x^2+4x+15x+5$

Factor out a common factor of x from the first two terms. Factor out a common factor of 3 from the second two terms.

$2x^2+3x+6x+9=x(2x+3)+3(2x+3)$

The expression in parentheses is now a common factor of the terms.

$x(2x+3)+3(2x+3)=(2x+3)(x+3)$

Vocabulary Answers: 1. common factor 2. factoring by grouping 3. commutative property 4. FOIL

Example	Student Practice
5. Factor. $4x + 8y + ax + 2ay$	**6.** Factor. $7y + 21z + xy + 3xz$

Factor out a common factor of 4 from the first two terms. Factor out a common factor of a from the second two terms.

$4x + 8y + ax + 2ay$
$= 4(x + 2y) + a(x + 2y)$

The common factor of the terms is the expression in parentheses, $x + 2y$.

$4(x + 2y) + a(x + 2y) = (x + 2y)(4 + a)$

7. Factor. $bx + 4y + 4b + xy$	**8.** Factor. $nz + 6y + 6z + ny$

Rearrange the terms so that the first two terms have a common factor.

$bx + 4y + 4b + xy = bx + 4b + xy + 4y$

Factor out the common factor b from the first two terms and the common factor y from the second two terms.

$bx + 4b + xy + 4y = b(x + 4) + y(x + 4)$
$\qquad\qquad\qquad\quad = (x + 4)(b + y)$

9. Factor. $2x^2 + 5x - 4x - 10$	**10.** Factor. $2x^2 + 3x - 6x - 9$

Factor out the common factor x from the first two terms and the common factor -2 from the second two terms.

$2x^2 + 5x - 4x - 10$
$= x(2x + 5) - 4x - 10$
$= x(2x + 5) - 2(2x + 5)$
$= (2x + 5)(x - 2)$

176

Example	Student Practice
11. Factor. $2ax - a - 2bx + b$	**12.** Factor. $5nz - n - 5mz + m$

Factor out the common factor a from the first two terms and the common factor $-b$ from the second two terms.

$$2ax - a - 2bx + b = a(2x - 1) - b(2x - 1)$$

Since the two resulting terms contain the same parenthetical expression, we can complete the factoring.

$$a(2x - 1) - b(2x - 1) = (2x - 1)(a - b)$$

---

13. Factor and check your answer.
   $8ad + 21bc - 6bd - 28ac$

Use the commutative property of addition to rearrange the order so the first two terms have a common factor.

$$8ad + 21bc - 6bd - 28ac$$
$$= 8ad - 6bd - 28ac + 21bc$$

Factor out the common factor $2d$ from the first two terms and the common factor $-7c$ from the last two terms.

$$= 2d(4a - 3b) - 7c(4a - 3b)$$

Factor out the common factor $(4a - 3b)$.

$$= (4a - 3b)(2d - 7c)$$

To check, we multiply the two binomials using the FOIL procedure.

$$(4a - 3b)(2d - 7c)$$
$$= 8ad - 28ac - 6bd + 21bc$$
$$= 8ad + 21bc - 6bd - 28ac$$

14. Factor and check your answer.
   $9wz + 35xy - 15xz - 21wy$

Extra Practice

1. Factor by grouping. Check you answer.

$$x(x+1) - 2(x+1)$$

2. Factor by grouping. Check you answer.

$$x^2 - 2x + 5x - 10$$

3. Factor by grouping. Check you answer.

$$6x^2 + 15x - 4x - 10$$

4. Factor by grouping. Check you answer.

$$12ax + 15ay + 8xb + 10by$$

Concept Check

Explain how you would factor the following polynomial.

$$10ax + b^2 + 2bx + 5ab$$

Chapter 6 Factoring
6.3 Factoring Trinomials of the Form $x^2 + bx + c$

Vocabulary

first terms   •   outer and inner terms   •   second terms   •   last terms

1. When multiplying an expression of the form $(x+m)(x+n)$, the product of the
   _____ in the factors produces the first term of the polynomial.

2. When multiplying an expression of the form $(x+m)(x+n)$, the sum of the
   _____ in the factors gives the coefficient of the middle term of the polynomial.

3. When multiplying an expression of the form $(x+m)(x+n)$, the sum of the products of
   the _____ in the factors produces the middle term of the polynomial.

4. When multiplying an expression of the form $(x+m)(x+n)$, the product of the
   _____ of the factors gives the last term of the polynomial.

Example	Student Practice
1. Factor. $x^2 + 7x + 12$ The answer will be of the form $(x+m)(x+n)$. We want to find the two numbers, m and n, that you can multiply to get 12 and add to get 7. The numbers are 3 and 4. $x^2 + 7x + 12 = (x+3)(x+4)$	**2.** Factor. $x^2 + 9x + 18$
3. Factor. $x^2 + 12x + 20$ We want two numbers that have a product of 20 and a sum of 12. The numbers are 10 and 2. $x^2 + 12x + 20 = (x+10)(x+2)$	**4.** Factor. $x^2 + 14x + 33$

Vocabulary Answers: 1. first terms 2. second terms 3. outer and inner terms 4. last terms

Example	Student Practice
5. Factor. $x^2 - 8x + 15$	**6.** Factor. $x^2 - 11x + 28$

We want two numbers that have a product of $+15$ and a sum of -8. They must be negative numbers since the sign of the middle term is negative and the sign of the last term is positive.

The sum $-5 + (-3)$ is -8 and the product $(-5)(-3)$ is $+15$.

$$x^2 - 8x + 15 = (x-5)(x-3)$$

Multiply using FOIL to check.

Example	Student Practice
7. Factor. $x^2 - 3x - 10$	**8.** Factor. $x^2 - 2x - 24$

We want two numbers whose product is -10 and whose sum is -3. The two numbers are -5 and $+2$.

$$x^2 - 3x - 10 = (x-5)(x+2)$$

Example	Student Practice
9. Factor and check your answer. $y^2 + 10y - 24$	**10.** Factor and check your answer. $z^2 + 12z - 28$

The two numbers whose product is -24 and whose sum is $+10$ are $+12$ and -2.

$$y^2 + 10y - 24 = (y+12)(y-2)$$

It is very easy to make a sign error in these problems. Make sure that you mentally check your answer by FOIL to obtain the original expression.

Check.

$$(y+12)(y-2) = y^2 - 2y + 12y - 24$$
$$= y^2 + 10y - 24$$

Example	Student Practice
11. Factor. $y^4 - 2y^2 - 35$	**12.** Factor. $x^4 - 3x^2 - 10$

11. Factor. $y^4 - 2y^2 - 35$

Notice that $y^4 = (y^2)(y^2)$. This will be the first term in each set of parentheses.

$(y^2 \quad)(y^2 \quad)$

The last term of the polynomial is negative. Thus, the signs of m and n will be different.

$(y^2 + \quad)(y^2 - \quad)$

Now think of factors of 35 whose difference is 2.

$(y^2 + 5)(y^2 - 7)$

Multiply using FOIL to check.

---

13. Factor. $3x^2 + 9x - 162$

14. Factor. $4x^2 - 16x - 84$

13. Factor. $3x^2 + 9x - 162$

First factor out the common factor 3 from each term of the polynomial.

$3x^2 + 9x - 162 = 3(x^2 + 3x - 54)$

Then factor the remaining polynomial.

$3(x^2 + 3x - 54) = 3(x - 6)(x + 9)$

The final answer is $3(x - 6)(x + 9)$. Be sure to include the 3.

Check.

$$3(x - 6)(x + 9) = 3(x^2 + 3x - 54)$$
$$= 3x^2 + 9x - 162$$

Example	Student Practice

15. Find a polynomial in factored form for the shaded area in the figure.

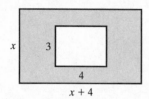

To obtain the shaded area, we find the area of the larger rectangle and subtract from it the area of the smaller rectangle. Thus, we have the following:

$$\text{shaded area} = x(x+4) - (4)(3)$$
$$= x^2 + 4x - 12$$

Now we factor this polynomial to obtain the shaded area $= (x+6)(x-2)$.

16. Find a polynomial in factored form for the shaded area in the figure.

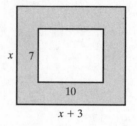

Extra Practice

1. Factor. $x^2 + 6x + 8$

2. Factor. $a^2 - 7a + 12$

3. Factor. $x^2 - x - 12$

4. Factor. $2x^2 + 18x + 28$

Concept Check

Explain how you would completely factor $4x^2 - 4x - 120$.

Chapter 6 Factoring
6.4 Factoring Trinomials of the Form $ax^2 + bx + c$

Vocabulary
trial-and-error method   •   grouping method   •   multiplying   •   greatest common factor

1. The _____ requires listing the possible factoring combinations and computing the middle terms by the FOIL method.

2. Always check by _____ the factors to see if the original trinomial is obtained.

3. First factor out the _____ if any, and then factor the trinomial.

4. The _____ for factoring trinomials of the form $ax^2 + bx + c$ requires writing the polynomial with four terms and then factoring the polynomial by grouping.

Example	**Student Practice**
1. Factor. $2x^2 + 5x + 3$	**2.** Factor. $2x^2 + 13x + 20$

Example

1. Factor. $2x^2 + 5x + 3$

In order for the coefficient of the x^2- term of the polynomial to be 2, the coefficients of the x-terms in the factors must be 2 and 1. Thus,

$$2x^2 + 5x + 3 = (2x \quad)(x \quad).$$

In order for the last term of the polynomial to be 3, the constants in the factors must be 3 and 1. Since all signs in the polynomial are positive, each factor in parentheses will be positive. We have two possibilities. We check them by multiplying.

$$(2x+1)(x+3) = 2x^2 + 7x + 3$$
$$(2x+3)(x+1) = 2x^2 + 5x + 3$$

The correct answer is $(2x+3)(x+1)$.

Vocabulary Answers: 1. trial-and-error method 2. multiplying 3. greatest common factor 4. grouping method

Example	Student Practice
3. Factor. $4x^2 - 13x + 3$	**4.** Factor. $3x^2 - 4x + 1$

3. Factor. $4x^2 - 13x + 3$

The different factorizations of 4 are $(2)(2)$ and $(1)(4)$. The factorization of 3 is $(1)(3)$. Let us list the possible factoring combinations and compute the middle term by the FOIL method. Note that the signs of the constants in both factors will be negative.

Possible Factors	Middle Term
$(2x-3)(2x-1)$	$-8x$
$(4x-3)(x-1)$	$-7x$
$(4x-1)(x-3)$	$-13x$

The correct middle term is $-13x$. The correct answer is $(4x-1)(x-3)$.

4. Factor. $3x^2 - 4x + 1$

---

5. Factor. $3x^2 - 2x - 8$

The factorization of 3 is $(1)(3)$. The factorizations of 8 are $(8)(1)$ and $(4)(2)$. Let us list only one-half of the possibilities.

Possible Factors	Middle Term
$(x+8)(3x-1)$	$+23x$
$(x+1)(3x-8)$	$-5x$
$(x+4)(3x-2)$	$+10x$
$(x+2)(3x-4)$	$+2x$

The middle term $+2x$ is only incorrect because the sign is wrong. So we just reverse the signs of the constraints.

The correct answer is $(x-2)(3x+4)$.

6. Factor. $4x^2 - 5x - 21$

Example	Student Practice

7. Factor by grouping. $2x^2 + 5x + 3$

First find the grouping number. The grouping number is $(2)(3) = 6$. The factors of 6 are $6 \cdot 1$ and $3 \cdot 2$. We choose the numbers 3 and 2 because their product is 6 and their sum is 5. Next, write $5x$ as the sum $2x + 3x$. Then, factor by grouping.

$$2x^2 + 5x + 3 = 2x^2 + 2x + 3x + 3$$
$$= 2x(x+1) + 3(x+1)$$
$$= (x+1)(2x+3)$$

Multiply to check.

$$(x+1)(2x+3) = 2x^2 + 3x + 2x + 3$$
$$= 2x^2 + 5x + 3$$

8. Factor by grouping. $2x^2 + 5x + 3$

9. Factor by grouping. $3x^2 - 2x - 8$

First find the grouping number. The grouping number is $(3)(-8) = -24$. Now we want two numbers whose product is -24 and whose sum is -2. They are -6 and 4. So we write $-2x$ as the sum $-6x + 4x$. Then, we factor by grouping.

$$3x^2 - 6x + 4x - 8 = 3x(x-2) + 4(x-2)$$
$$= (x-2)(3x+4)$$

10. Factor by grouping. $5x^2 - 13x - 6$

Example	Student Practice
11. Factor. $9x^2 + 3x - 30$ We first factor out the common factor 3 from each term of the trinomial. $9x^2 + 3x - 30 = 3(3x^2 + 1x - 10)$ We then factor the trinomial by the grouping method or by the trial-and-error method. $3(3x^2 + 1x - 10) = 3(3x - 5)(x + 2)$	**12.** Factor. $8x^2 + 22x - 6$
13. Factor. $32x^2 - 40x + 12$ We first factor out the greatest common factor 4 from each term of the trinomial. $32x^2 - 40x + 12 = 4(8x^2 - 10x + 3)$ We then factor the trinomial by the grouping method or by the trial-and-error method. $4(8x^2 - 10x + 3) = 4(2x - 1)(4x - 3)$	**14.** Factor. $30x^2 - 65x + 30$

Extra Practice

1. Factor by the trial-and-error method. Check your answer by using FOIL.
$4x^2 - 12x + 5$

2. Factor by the grouping method. Check your answer by using FOIL. $7x^2 - 4x - 3$

3. Factor by any method. $5x^2 + 43x - 18$

4. Factor by first factoring out the greatest common factor. $15y^2 + 51y - 36$

Concept Check
Explain how you would factor $10x^3 + 18x^2y - 4xy^2$.

Chapter 6 Factoring
6.5 Special Cases of Factoring

Vocabulary
difference of two squares • negative • perfect-square trinomials • greatest common factor

1. The difference of two squares formula only works if the last term is _____.

2. A _____ is a trinomial where the first and last terms are perfect squares and the middle term is twice the product of the values whose squares are the first and last terms.

3. The _____ can be factored into the sum and difference of those values that were squared.

4. For some polynomials, we first need to factor out the _____.

Example	Student Practice
1. Factor. $9x^2 - 1$	**2.** Factor. $36x^2 - 1$
We see that the polynomial is in the form of the difference of two squares. $9x^2$ is a square and 1 is a square. $9x^2 = (3x)^2$ and $1 = (1)^2$. So using the formula we can write the following. $$9x^2 - 1 = (3x+1)(3x-1)$$	
3. Factor. $25x^2 - 16$	**4.** Factor. $64x^2 - 9$
$25x^2 = (5x)^2$ and $16 = (4)^2$. Again we use the formula for the difference of two squares. $$25x^2 - 16 = (5x+4)(5x-4)$$	

Vocabulary Answers: 1. negative 2. perfect-square trinomials 3. difference of two squares 4. greatest common factor

Example	Student Practice
5. Factor. $81x^4 - 1$	**6.** Factor. $16x^4 - 1$

Because $81x^4 = \left(9x^2\right)^2$ and $1 = \left(1\right)^2$, we see that

$$81x^4 - 1 = \left(9x^2 + 1\right)\left(9x^2 - 1\right).$$

The factoring is not complete. We can factor $9x^2 - 1$, because
$$9x^2 - 1 = \left(3x + 1\right)\left(3x - 1\right).$$

$$81x^4 - 1 = \left(9x^2 + 1\right)\left(3x + 1\right)\left(3x - 1\right)$$

7. Factor. $x^2 + 6x + 9$	**8.** Factor. $x^2 + 10x + 25$

This is a perfect-square trinomial. The first and last terms are perfect squares because $x^2 = \left(x\right)^2$ and $9 = \left(3\right)^2$. The middle term, $6x$, is twice the product of 3 and x.

Since $x^2 + 6x + 9$ is a perfect-square trinomial we can use the formula

$$a^2 + 2ab + b^2 = \left(a + b\right)^2$$

with $a = x$ and $b = 3$. So we have

$$x^2 + 6x + 9 = \left(x + 3\right)^2.$$

Example	Student Practice
9. Factor.	**10.** Factor.

(a) $49x^2 + 42xy + 9y^2$

This is a perfect square trinomial, because $49x^2 = (7x)^2$, $9y^2 = (3y)^2$, and $42xy = 2(7x \cdot 3y)$.

$$49x^2 + 42xy + 9y^2 = (7x + 3y)^2$$

(b) $36x^4 - 12x^2 + 1$

This is a perfect square trinomial, because $36x^4 = (6x^2)^2$, $1 = (1)^2$, and $12x^2 = 2(6x^2 \cdot 1)$.

$$36x^4 - 12x^2 + 1 = (6x^2 - 1)^2$$

(a) $64x^2 + 80xy + 25y^2$

(b) $49x^4 - 14x^2 + 1$

---

11. Factor. $49x^2 + 35x + 4$

This is not a perfect-square trinomial! Although the first and last terms are perfect squares since $(7x)^2 = 49x^2$ and $(2)^2 = 4$, the middle term, $35x$, is not double the product of 2 and $7x$. $35x \neq 28x$! So we must factor by trial and error or by grouping to obtain

$$49x^2 + 35x + 4 = (7x + 4)(7x + 1).$$

12. Factor. $36x^2 + 25x + 4$

Example	Student Practice
13. Factor. $12x^2 - 48$	**14.** Factor. $3x^2 - 75$

We see that the greatest common factor is 12. First we factor out 12. Then we use the difference-of-two-squares formula, $a^2 - b^2 = (a+b)(a-b)$.

$$12x^2 - 48 = 12(x^2 - 4)$$
$$= 12(x+2)(x-2)$$

15. Factor. $24x^2 - 72x + 54$ **16.** Factor. $48x^2 - 72x + 27$

First we factor out the greatest common factor, 6. Then we use the perfect-square-trinomial formula,
$$a^2 - 2ab + b^2 = (a-b)^2.$$

$$24x^2 - 72x + 54 = 6(4x^2 - 12x + 9)$$
$$= 6(2x-3)^2$$

Extra Practice

1. Factor using the difference-of-two-squares formula. $25a^2 - 36b^2$

2. Factor by using the perfect-square trinomial formula. $36a^2 + 60ab + 25b^2$

3. Factor by using the difference-of-two-squares. $x^4 - 100$

4. Factor by using the perfect-square trinomial formula. $9x^4 - 30x^2 + 25$

Concept Check

Explain how to factor the polynomial $24x^2 + 120x + 150$.

Chapter 6 Factoring
6.6 A Brief Review of Factoring

Vocabulary

perfect-square trinomial • difference of two squares • factor by grouping • prime

trinomial of the form $x^2 + bx + c$ • trinomial of the form $ax^2 + bx + c$ • common factor

1. Some polynomials have a _____ consisting of a number, a variable, or both.

2. When we _____, we rearrange the order if the first two terms do not have a common factor.

3. If we cannot factor a polynomial by elementary methods, we will identify it as a _____ polynomial.

4. In a _____ there are three terms and the first and last terms are perfect squares.

Example	Student Practice
1. Factor.	**2.** Factor.
(a) $25x^3 - 10x^2 + x$	**(a)** $9y^3 + 6y^2 + y$
Factor out the common factor x. The other factor is a perfect-square trinomial.	
$$25x^3 - 10x^2 + x = x(25x^2 - 10x + 1)$$ $$= x(5x - 1)^2$$	
(b) $20x^2y^2 - 45y^2$	**(b)** $-4x^3 + 28x^2 + 32x$
Factor out the common factor $5y^2$. The other factor is a difference of squares.	
$$20x^2y^2 - 45y^2 = 5y^2(4x^2 - 9)$$ $$= 5y^2(2x + 3)(2x - 3)$$	

Vocabulary Answers: 1. common factor 2. factor by grouping 3. prime 4. perfect-square trinomial

Example	Student Practice
3. Factor. $ax^2 - 9a + 2x^2 - 18$	**4.** Factor. $bx^2 - 16b - 3x^2 + 48$

3. Factor. $ax^2 - 9a + 2x^2 - 18$

We factor by grouping since there are four terms. Factor out the common factor a from the first two terms and 2 from the second two terms.

$$ax^2 - 9a + 2x^2 - 18$$
$$= a\left(x^2 - 9\right) + 2\left(x^2 - 9\right)$$

Factor out the common factor $\left(x^2 - 9\right)$.

$$a\left(x^2 - 9\right) + 2\left(x^2 - 9\right) = (a + 2)\left(x^2 - 9\right)$$

Factor $x^2 - 9$ using the difference-of-two squares formula.

$$(a + 2)\left(x^2 - 9\right) = (a + 2)(x - 3)(x + 3)$$

4. Factor. $bx^2 - 16b - 3x^2 + 48$

---

5. Factor, if possible.

(a) $x^2 + 6x + 12$

The factors of 12 are

$$(1)(12) \text{ or } (2)(6) \text{ or } (3)(4).$$

None of these pairs add up to 6, the coefficient of the middle term. Thus, the problem cannot be factored by the methods in this chapter.

(b) $25x^2 + 4$

We have a formula to factor the difference of two squares. There is no way to factor the sum of two squares. That is, $a^2 + b^2$ cannot be factored.

6. Factor, if possible.

(a) $x^2 + 5x + 17$

(b) $x^2 - x + 21$

Extra Practice

1. Factor, if possible. Be sure to factor completely. Factor out the greatest common factor first, if one exists.

$x^2 + 9$

2. Factor, if possible. Be sure to factor completely. Factor out the greatest common factor first, if one exists.

$63x - 7x^3$

3. Factor, if possible. Be sure to factor completely. Factor out the greatest common factor first, if one exists.

$-x^3 + 12x^2 + 45x$

4. Factor, if possible. Be sure to factor completely. Factor out the greatest common factor first, if one exists.

$x^2 + 2x + 3x + 6$

Concept Check

Explain how to completely factor $2x^2 + 6xw - 5x - 15w$.

Chapter 6 Factoring
6.7 Solving Quadratic Equations by Factoring

Vocabulary
quadratic equation　•　standard form　•　real roots　•　zero factor property

1. Many quadratic equations have two real number solutions, also called _____.

2. The _____ states that if $a \cdot b = 0$, then $a = 0$ or $b = 0$.

3. A _____ is a polynomial equation in one variable that contains a variable term of degree 2 and no terms of higher degree.

Example	Student Practice
1. Solve the equation to find the two roots. $2x^2 + 13x - 7 = 0$	**2.** Solve the equation to find the two roots. $3x^2 + 8x - 3 = 0$

The equation is in standard form. Factor. Then set each factor equal to 0 and solve the equations to find the two roots.

$$2x^2 + 13x - 7 = 0$$
$$(2x - 1)(x + 7) = 0$$
$$2x - 1 = 0 \quad x + 7 = 0$$
$$x = \frac{1}{2} \quad\quad x = -7$$

Check. If $x = \frac{1}{2}$ and if $x = -7$ then we have the following.

$$2\left(\frac{1}{2}\right)^2 + 13\left(\frac{1}{2}\right) - 7 = \frac{1}{2} + \frac{13}{2} - \frac{14}{2} = 0$$
$$2(-7)^2 + 13(-7) - 7 = 98 - 91 - 7 = 0$$

Thus $\frac{1}{2}$ and -7 are both roots of the equation $2x^2 + 13x - 7 = 0$.

Vocabulary Answers: 1. real roots 2. zero factor property 3. quadratic equation

Example	Student Practice
3. Solve the equation to find the two roots. $7x^2 - 3x = 0$	**4.** Solve the equation to find the two roots. $5x^2 - 4x = 0$

3. Solve the equation to find the two roots.
$7x^2 - 3x = 0$

The equation is in standard form. Here $c = 0$. Factor out the common factor. Then set each factor equal to 0 by the zero factor property. Solve the equations to find the two roots.

$$7x^2 - 3x = 0$$
$$x(7x - 3) = 0$$
$$x = 0 \quad 7x - 3 = 0$$
$$7x = 3$$
$$x = \frac{3}{7}$$

The two roots are 0 and $\frac{3}{7}$.

Check. Verify that 0 and $\frac{3}{7}$ are the roots of $7x^2 - 3x = 0$.

4. Solve the equation to find the two roots.
$5x^2 - 4x = 0$

---

5. Solve. $x^2 = 12 - x$

The equation is not in standard form. Add x and -12 to both sides of the equation so that the left side is equal to zero. Then factor and set each factor equal to 0. Solve the equations for x.

$$x^2 = 12 - x$$
$$x^2 + x - 12 = 0$$
$$(x - 3)(x + 4) = 0$$
$$x - 3 = 0 \quad x + 4 = 0$$
$$x = 3 \qquad x = -4$$

The check is left to the student.

6. Solve. $x^2 = 63 - 2x$

Example	Student Practice

7. Carlos lives in Mexico City. He has a rectangular brick walkway in front of his house. The length of the walkway is 3 meters longer than twice the width. The area of the walkway is 44 square meters. Find the length and width of the rectangular walkway.

Let w = the width in meters.

Then $2w + 3$ = the length in meters.

Next, write an equation.

$$\text{area} = (\text{width})(\text{length})$$
$$44 = w(2w + 3)$$

Now, solve and state the answer.

$$44 = w(2w + 3)$$
$$44 = 2w^2 + 3w$$
$$0 = 2w^2 + 3w - 44$$
$$0 = (2w + 11)(w - 4)$$
$$2w + 11 = 0 \qquad w - 4 = 0$$
$$w = -5\frac{1}{2} \qquad w = 4$$

Since it would not make sense to have a rectangle with a negative number as a width, $-5\frac{1}{2}$ is not a valid solution.

Since $w = 4$, the width of the walkway is 4 meters. The length is $2w + 3$, so we have $2(4) + 3 = 8 + 3 = 11$. Thus, the length of the walkway is 11 meters.

The check is left to the student.

8. The length of a rectangle is 31 inches shorter than triple the width. The rectangle has an area of 60 square inches. Find the length and width of the rectangle.

Example	Student Practice
9. A tennis ball is thrown upward with an initial velocity of 8 meters/second. Suppose the initial height above the ground is 4 meters. At what time t will the ball hit the ground?	**10.** A baseball is thrown upward with an initial velocity of 9 meters/second. Suppose the initial height above the ground is 18 meters. At what time t will the ball hit the ground?

In this case $S = 0$ since the ball will hit the ground. The initial upward velocity is $v = 8$ meters/second. The initial height is 4 meters, so $h = 4$.

$$S = -5t^2 + vt + h$$
$$0 = -5t^2 + 8t + 4$$
$$5t^2 - 8t - 4 = 0$$
$$(5t + 2)(t - 2) = 0$$
$$5t + 2 = 0 \qquad t - 2 = 0$$
$$t = -\frac{2}{5} \qquad t = 2$$

We want a positive time for t in seconds; thus we do not use $t = -\frac{2}{5}$.

Therefore, the ball will strike the ground 2 seconds after it is thrown.

The check is left to the student.

Extra Practice

1. Solve for the roots of the quadratic equation. Check your answer.
$2x^2 - 5x - 3 = 0$

2. Solve for the roots of the quadratic equation. Check your answer.
$3x^2 = 27$

3. Solve for the roots of the quadratic equation. Check your answer.
$x^2 - 35 = 2x$

4. Solve for the roots of the quadratic equation. Check your answer.
$x^2 - 7x = -12$

Concept Check

Explain how you would solve the following problem: A rectangle has an area of 65 square feet. The length of the rectangle is 3 feet longer than double the width. Find the length and the width of the rectangle.

MATH COACH

Mastering the skills you need to do well on the test.

Watch the MATH COACH videos in MyMathLab® or on You Tube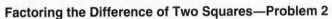 while you work the problems below. These helpful hints will help you avoid making common errors on test problems.

Factoring the Difference of Two Squares—Problem 2

Factor completely. $16x^2 - 81$

> **Helpful Hint:** It is important to learn the difference-of-two-squares formula: $a^2 - b^2 = (a+b)(a-b)$. Remember that the numerical values in both terms will be perfect squares. The first ten perfect squares are 1, 4, 9, 16, 25, 36, 49, 64, 81, and 100.

Did you remember that $(4x)^2 = 16x^2$? Yes ____ No ____

Did you remember that $9^2 = 81$? Yes ____ No ____

If you answered No to these questions, stop and review the list of the first ten perfect squares. Consider that $4^2 = 16$ and $x \cdot x = x^2$.

If you answered No, stop and review the difference-of-two-squares formula again. Make sure that one set of parentheses contains a + sign and the other set of parentheses contains a − sign.

If you answered Problem 2 incorrectly, go back and rework the problem using these suggestions.

Do you see how $16x^2 - 81$ can be factored using the formula $a^2 - b^2$? Yes ____ No ____

Factoring a Perfect-Square Trinomial—Problem 4   Factor completely. $9a^2 - 30a + 25$

> **Helpful Hint:** Remember the perfect-square-trinomial formula: $a^2 - 2ab + b^2 = (a-b)^2$. You must verify two things to determine if you can use this formula:
> (1) The numerical values in the first term and the last term must be perfect squares.
> (2) The middle term must equal "twice the product of the values whose squares are the first and last terms."

Did you remember that $(3a)^2 = 9a^2$? Yes ____ No ____

Did you remember that $5^2 = 25$? Yes ____ No ____

If you answered No to these questions, stop and review the first ten perfect squares. Consider that $3^2 = 9$ and $a \cdot a = a^2$.

If you answered No, check to see if the middle term, $30a$, equals twice the product of $3a$ and 5.

Now go back and rework the problem using these suggestions.

Do you see how $9a^2 - 30a + 25$ can be factored using the formula $(a-b)^2$? Yes ____ No ____

Factoring a Polynomial with Four Terms by Grouping—Problem 6

Factor completely. $10xy + 15by - 8x - 12b$

> **Helpful Hint:** Look for common factors first. We can find the greatest common factors of the first two terms and factor. Then we can find the greatest common factor of the second two terms and factor. Make sure that you obtain the same binomial factor for each step. Be careful with $+/-$ signs.

Did you identify $5y$ as the greatest common factor of the first two terms: $10xy + 15by$?
Yes ____ No ____

If you answered No, remember that to factor completely, we must remove all common factors from both terms. Stop now and complete this step.

Did you identify 4 as the greatest common factor of the second two terms: $-8x - 12b$?
Yes ____ No ____

If you answered Problem 6 incorrectly, go back and rework the problem using these suggestions.

If you answered Yes to these questions, then you obtained $(2x + 3b)$ in the first term and $(-2x - 3b)$ in the second term. These are not the same binomial factor. Stop and consider how to get the same binomial factor of $(2x + 3b)$.

In your final answer, is the binomial factor of $(2x + 3b)$ listed once?
Yes ____ No ____

Factoring a Polynomial with a Common Factor—Problem 19   Factor completely. $3x^2 - 3x - 90$

> **Helpful Hint:** Look for the greatest common factor of all three terms as your first step. Don't forget to include this common factor as part of your answer. Always check your final product to make sure that it matches the original polynomial. Do this by multiplying.

Did you obtain $(3x - 18)(x + 5)$ or $(3x - 15)(x - 6)$ as your answer?
Yes ____ No ____

Now go back and rework the problem using these suggestions.

If you answered Yes, then you forgot to factor out the greatest common factor 3 as your first step.

Do you see how to factor $x^2 - x - 30$?
Yes ____ No ____

If you answered No, remember that we are looking for two numbers with a product of -30 and a sum of -1.

Be sure to double-check your final answer to be sure there are no common factors and include 3 in your final answer.

Chapter 7 Rational Expressions and Equations
7.1 Simplifying Rational Expressions

Vocabulary
rational expression • fractional algebraic expression • basic rule of fractions
simplify the fraction • factors

1. Only _____ of both the numerator and the denominator can be divided out.

2. A _____ is a polynomial divided by another polynomial.

3. Dividing out common factors is how to _____.

4. The _____ states that for any rational expression $\dfrac{a}{b}$ and any polynomials a, b,

   and c, $\dfrac{ac}{bc} = \dfrac{a}{b}$.

Example	Student Practice
1. Reduce. $\dfrac{21}{39}$	**2.** Reduce. $\dfrac{56}{77}$
Use the rule $\dfrac{ac}{bc} = \dfrac{a}{b}$. Let $c = 3$.	
$\dfrac{21}{39} = \dfrac{7 \cdot \cancel{3}}{13 \cdot \cancel{3}} = \dfrac{7}{13}$	
3. Simplify. $\dfrac{4x+12}{5x+15}$	**4.** Simplify. $\dfrac{6x+8}{15x+20}$
$\dfrac{4x+12}{5x+15} = \dfrac{4(x+3)}{5(x+3)}$ $= \dfrac{4\cancel{(x+3)}}{5\cancel{(x+3)}}$ $= \dfrac{4}{5}$	

Vocabulary Answers: 1. factors 2. rational expression 3. simplify the fraction 4. basic rule of fractions

Example	Student Practice
5. Simplify. $\dfrac{x^2+9x+14}{x^2-4}$	**6.** Simplify. $\dfrac{x^2+3x-28}{x^2-16}$

$$\frac{x^2+9x+14}{x^2-4} = \frac{(x+7)(x+2)}{(x-2)(x+2)}$$

$$= \frac{(x+7)\,\cancel{(x+2)}}{(x-2)\,\cancel{(x+2)}}$$

$$= \frac{x+7}{x-2}$$

7. Simplify. $\dfrac{x^3-9x}{x^3+x^2-6x}$	**8.** Simplify. $\dfrac{x^3+8x^2+12x}{x^3-36x}$

$$\frac{x^3-9x}{x^3+x^2-6x} = \frac{x\left(x^2-9x\right)}{x\left(x^2+x-6\right)}$$

$$= \frac{\cancel{x}\,\cancel{(x+3)}(x-3)}{\cancel{x}\,\cancel{(x+3)}(x-2)}$$

$$= \frac{x-3}{x-2}$$

9. Simplify. $\dfrac{5x-15}{6-2x}$	**10.** Simplify. $\dfrac{8x-24}{42-14x}$

The variable terms in the numerator and in the denominator are opposite in sign. Likewise the numerical terms are opposite in sign. Factor out a negative number from the denominator.

$$\frac{5x-15}{6-2x} = \frac{5(x-3)}{-2(-3+x)}$$

$$= \frac{5\cancel{(x-3)}}{-2\cancel{(-3+x)}}$$

$$= -\frac{5}{2}$$

Example	Student Practice

11. Simplify. $\dfrac{2x^2 - 11x + 12}{16 - x^2}$

Factor the numerator and the denominator. Observe that $(x-4)$ and $(4-x)$ are opposites.

$$\frac{2x^2 - 11x + 12}{16 - x^2} = \frac{(x-4)(2x-3)}{(4-x)(4+x)}$$

Factor -1 out of $(+4-x)$ to obtain $-1(-4+x)$.

$$\frac{(x-4)(2x-3)}{(4-x)(4+x)} = \frac{(x-4)(2x-3)}{-1(-4+x)(4+x)}$$

$$= \frac{(x-4)(2x-3)}{-1(-4+x)(4+x)}$$

$$= \frac{(2x-3)}{-1(4+x)}$$

$$= -\frac{2x-3}{4+x}$$

12. Simplify. $\dfrac{6x^2 - 7x - 20}{25 - 4x^2}$

13. Simplify. $\dfrac{x^2 - 7xy + 12y^2}{2x^2 - 7xy - 4y^2}$

$$\frac{x^2 - 7xy + 12y^2}{2x^2 - 7xy - 4y^2} = \frac{(x-4y)(x-3y)}{(2x+y)(x-4y)}$$

$$= \frac{(x-4y)(x-3y)}{(2x+y)(x-4y)}$$

$$= \frac{x-3y}{2x+y}$$

14. Simplify. $\dfrac{10x^2 - 9xy - 9y^2}{18x^2 - 13xy - 21y^2}$

Example	Student Practice
15. Simplify. $\dfrac{6a^2 + ab - 7b^2}{36a^2 - 49b^2}$	**16.** Simplify. $\dfrac{25a^2 - 16b^2}{10a^2 - 3ab - 4b^2}$

$$\frac{6a^2 + ab - 7b^2}{36a^2 - 49b^2} = \frac{(6a + 7b)(a - b)}{(6a + 7b)(6a - 7b)}$$

$$= \frac{\cancel{(6a + 7b)}(a - b)}{\cancel{(6a + 7b)}(6a - 7b)}$$

$$= \frac{a - b}{6a - 7b}$$

Extra Practice

1. Simplify. $\dfrac{3x + 12}{x^2 + 4x}$

2. Simplify. $\dfrac{9x^2 - 24x + 16}{9x^2 - 16}$

3. Simplify. $\dfrac{81 - x^2}{3x^2 - 21x - 54}$

4. Simplify. $\dfrac{25x^2 - 20xy + 4y^2}{15x^2 - xy - 2y^2}$

Concept Check

Explain why it is important to completely factor both the numerator and the denominator when simplifying $\dfrac{x^2y - y^3}{x^2y + xy^2 - 2y^3}$.

Chapter 7 Rational Expressions and Equations
7.2 Multiplying and Dividing Rational Expressions

Vocabulary

reciprocals • multiply • divide • greatest common factor

1. You should always check for the _____ as your first step when multiplying rational expressions.

2. Two numbers are _____ of each other if their product is 1.

3. To _____ two rational expressions, multiply the numerators and multiply the denominators.

4. To _____ two rational expressions, invert the second fraction and multiply it by the first fraction.

Example	**Student Practice**
1. Multiply. $\dfrac{x^2-x-12}{x^2-16}\cdot\dfrac{2x^2+7x-4}{x^2-4x-21}$	**2.** Multiply. $\dfrac{3x^2+34x+63}{2x^2+7x-15}\cdot\dfrac{x^2+9x+20}{x^2+13x+36}$

Factoring is always the first step.

$$\frac{x^2-x-12}{x^2-16}\cdot\frac{2x^2+7x-4}{x^2-4x-21}$$
$$=\frac{(x-4)(x+3)}{(x-4)(x+4)}\cdot\frac{(x+4)(2x-1)}{(x+3)(x-7)}$$

Apply the basic rule of fractions. Three pairs of factors divide out.

$$\frac{(x-4)(x+3)}{(x-4)(x+4)}\cdot\frac{(x+4)(2x-1)}{(x+3)(x-7)}$$
$$=\frac{\cancel{(x-4)}\,\cancel{(x+3)}}{\cancel{(x-4)}\,\cancel{(x+4)}}\cdot\frac{\cancel{(x+4)}\,(2x-1)}{\cancel{(x+3)}\,(x-7)}$$
$$=\frac{(2x-1)}{(x-7)}$$

Vocabulary Answers: 1. greatest common factor 2. reciprocals 3. multiply 4. divide

Example	Student Practice

3. Multiply. $\dfrac{x^4 - 16}{x^3 + 4x} \cdot \dfrac{2x^2 - 8x}{4x^2 + 2x - 12}$

Factor each numerator and denominator. Factoring out the greatest common factor first is very important.

$\dfrac{x^4 - 16}{x^3 + 4x} \cdot \dfrac{2x^2 - 8x}{4x^2 + 2x - 12}$

$= \dfrac{\left(x^2 + 4\right)\left(x^2 - 4\right)}{x\left(x^2 + 4\right)} \cdot \dfrac{2x(x-4)}{2\left(2x^2 + x - 6\right)}$

$= \dfrac{\left(x^2 + 4\right)(x+2)(x-2)}{x\left(x^2 + 4\right)} \cdot \dfrac{2x(x-4)}{2(x+2)(2x-3)}$

$= \dfrac{\cancel{\left(x^2+4\right)}\cancel{(x+2)}(x-2)}{\cancel{x}\cancel{\left(x^2+4\right)}} \cdot \dfrac{\cancel{2}\cancel{x}(x-4)}{\cancel{2}\cancel{(x+2)}(2x-3)}$

$= \dfrac{(x-2)(x-4)}{2x-3}$ or $\dfrac{x^2 - 6x + 8}{2x-3}$

4. Multiply. $\dfrac{x^3 + x}{2x^4 - 2} \cdot \dfrac{4x^3 - 4x}{6x^3 + 38x^2 + 40x}$

---

5. Divide. $\dfrac{6x + 12y}{2x - 6y} \div \dfrac{9x^2 - 36y^2}{4x^2 - 36y^2}$

$= \dfrac{6x + 12y}{2x - 6y} \cdot \dfrac{4x^2 - 36y^2}{9x^2 - 36y^2}$

$= \dfrac{6(x + 2y)}{2(x - 3y)} \cdot \dfrac{4\left(x^2 - 9y^2\right)}{9\left(x^2 - 4y^2\right)}$

$= \dfrac{(3)(2)(x+2y)}{2(x-3y)} \cdot \dfrac{(2)(2)(x+3y)(x-3y)}{(3)(3)(x+2y)(x-2y)}$

$= \dfrac{\cancel{(3)}\cancel{(2)}\cancel{(x+2y)}}{\cancel{2}\cancel{(x-3y)}} \cdot \dfrac{(2)(2)(x+3y)\cancel{(x-3y)}}{\cancel{(3)}(3)\cancel{(x+2y)}(x-2y)}$

$= \dfrac{(2)(2)(x+3y)}{3(x-2y)}$

$= \dfrac{4(x+3y)}{3(x-2y)}$

6. Divide.

$\dfrac{5x^2 + 20x - 105}{4x^2 - 44x + 96} \div \dfrac{10x^2 + 75x + 35}{16x^2 - 140x + 96}$

Example	Student Practice

7. Divide. $\dfrac{15-3x}{x+6} \div \left(x^2 - 9x + 20\right)$

Note that $x^2 - 9x + 20$ can be written as $\dfrac{x^2 - 9x + 20}{1}$.

$$\dfrac{15-3x}{x+6} \div \left(x^2 - 9x + 20\right)$$

$$= \dfrac{15-3x}{x+6} \cdot \dfrac{1}{x^2 - 9x + 20}$$

$$= \dfrac{-3(-5+x)}{x+6} \cdot \dfrac{1}{(x-5)(x-4)}$$

$$= \dfrac{-3(\cancel{-5+x})}{x+6} \cdot \dfrac{1}{\cancel{(x-5)}(x-4)}$$

$$= \dfrac{-3}{(x+6)(x-4)}$$

or $-\dfrac{3}{(x+6)(x-4)}$ or $\dfrac{3}{(x+6)(4-x)}$

8. Divide. $\dfrac{x+4}{x-2} \div \left(-x^2 + x + 20\right)$

Extra Practice

1. Multiply. $\dfrac{32x^3}{8x^2 - 8} \cdot \dfrac{4x-4}{16x^2}$

2. Multiply.

$$\dfrac{2x^2 - 3x}{4x^3 + 4x^2 - 9x - 9} \cdot \dfrac{2x^2 + 5x + 3}{5x + 7}$$

3. Divide. $\dfrac{9x^2 - 49}{3x^2 + 8x - 35} \div \left(3x^2 + x - 14\right)$

4. Divide. $\dfrac{9x^2 - 16}{9x^2 + 24x + 16} \div \dfrac{3x^2 - x - 4}{5x^2 - 5}$

Concept Check

Explain how you would divide $\dfrac{21x - 7}{9x^2 - 1} \div \dfrac{1}{3x + 1}$.

Chapter 7 Rational Expressions and Equations
7.3 Adding and Subtracting Rational Expressions

Vocabulary

least common denominator   •   denominator   •   different   •   factor

1. If rational expressions have the same _____, they can be combined in the same way as arithmetic fractions.

2. If two rational expressions have _____ denominators, we first change them to equivalent rational expressions with the least common denominator.

3. The _____ is a product containing each different factor for each denominator of rational expressions.

4. The first step to find the LCD of two or more rational expressions is to _____ each denominator completely.

Example	**Student Practice**
1. Add. $\dfrac{5a}{a+2b}+\dfrac{6a}{a+2b}$	**2.** Add. $\dfrac{4x+3}{3x+4}+\dfrac{5-4x}{3x+4}$
Note that the denominators are the same. Only add the numerators. Do not change the denominator.	
$\dfrac{5a}{a+2b}+\dfrac{6a}{a+2b}=\dfrac{5a+6a}{a+2b}=\dfrac{11a}{a+2b}$	
3. Subtract. $\dfrac{3x}{(x+y)(x-2y)}-\dfrac{8x}{(x+y)(x-2y)}$	**4.** Subtract. $\dfrac{4x+6}{(3x-2)(x+9)}-\dfrac{2x-5}{(3x-2)(x+9)}$
Write as one fraction and simplify.	
$\dfrac{3x}{(x+y)(x-2y)}-\dfrac{8x}{(x+y)(x-2y)}$ $=\dfrac{3x-8x}{(x+y)(x-2y)}=\dfrac{-5x}{(x+y)(x-2y)}$	

Vocabulary Answers: 1. denominator 2. different 3. least common denominator 4. factor

Example	Student Practice
5. Find the LCD. $\dfrac{5}{2x-4}, \dfrac{6}{3x-6}$	**6.** Find the LCD. $\dfrac{5}{16x+20}, \dfrac{11}{28x+35}$

5. (continued)

Factor each denominator.

$$2x-4 = 2(x-2)$$
$$3x-6 = 3(x-2)$$

The factors are 2, 3, and $(x-2)$. The LCD is the product of these factors.

$$LCD = (2)(3)(x-2) = 6(x-2)$$

---

7. Find the LCD.

$$\frac{5}{12ab^2c}, \frac{13}{18a^3bc^4}$$

The LCD will contain each factor repeated the greatest number of times that it occurs in any one denominator.

$$12ab^2x = 2\cdot2\cdot3 \quad\cdot a \qquad \cdot b\cdot b\cdot c$$
$$18a^3bc^4 = \quad 2\cdot3\cdot3\cdot a\cdot a\cdot a\cdot b\cdot \; c\cdot c\cdot c\cdot c$$
$$LCD = 2\cdot2\cdot3\cdot3\cdot a\cdot a\cdot a\cdot b\cdot b\cdot c\cdot c\cdot c\cdot c$$
$$LCD = 2^2\cdot3^2\cdot a^3\cdot b^2\cdot c^4 = 36a^3b^2c^4$$

8. Find the LCD.

$$\frac{7}{24ab^2c^3}, \frac{13}{36a^2b^4c}$$

---

9. Add. $\dfrac{5}{xy}+\dfrac{2}{y}$

Find the LCD. The two factors are x and y. Observe that the LCD is xy.

$$\frac{5}{xy}+\frac{2}{y} = \frac{5}{xy}+\frac{2}{y}\cdot\frac{x}{x}$$
$$= \frac{5}{xy}+\frac{2x}{xy}$$
$$= \frac{5+2x}{xy}$$

10. Add. $\dfrac{4}{xyz}+\dfrac{2}{y}$

Example	Student Practice

11. Add. $\dfrac{3x}{x^2 - y^2} + \dfrac{5}{x+y}$

Factor the first denominator so that $x^2 - y^2 = (x+y)(x-y)$. Thus, the factors of the denominators are $(x+y)$ and $(x-y)$. Observe that the LCD $= (x+y)(x-y)$.

$\dfrac{3x}{x^2 - y^2} + \dfrac{5}{x+y}$

$= \dfrac{3x}{(x+y)(x-y)} + \dfrac{5}{(x+y)} \cdot \dfrac{x-y}{x-y}$

$= \dfrac{3x}{(x+y)(x-y)} + \dfrac{5x-5y}{(x+y)(x-y)}$

$= \dfrac{8x-5y}{(x+y)(x-y)}$

12. Add. $\dfrac{4}{2x-5} + \dfrac{3x+2}{4x^2 - 25}$

---

13. Add. $\dfrac{5}{x^2 - y^2} + \dfrac{3x}{x^3 + x^2 y}$

$\dfrac{5}{x^2 - y^2} + \dfrac{3x}{x^3 + x^2 y}$

$= \dfrac{5}{(x+y)(x-y)} + \dfrac{3x}{x^2(x+y)}$

$= \dfrac{5}{(x+y)(x-y)} \cdot \dfrac{x^2}{x^2} + \dfrac{3x}{x^2(x+y)} \cdot \dfrac{x-y}{x-y}$

$= \dfrac{5x^2}{x^2(x+y)(x-y)} + \dfrac{3x^2 - 3xy}{x^2(x+y)(x-y)}$

$= \dfrac{x(8x-3y)}{x^2(x+y)(x-y)}$

$= \dfrac{8x-3y}{x(x+y)(x-y)}$

14. Add. $\dfrac{6y}{3xy - y^2} + \dfrac{4y}{9x^2 - y^2}$

Example	Student Practice
15. Subtract. $\dfrac{3x+4}{x-2} - \dfrac{x-3}{2x-4}$	**16.** Subtract. $\dfrac{x+5}{2x+7} - \dfrac{x-2}{6x+21}$

Factor the second denominator.

$$\frac{3x+4}{x-2} - \frac{x-3}{2x-4} = \frac{3x+4}{x-2} - \frac{x-3}{2(x-2)}$$

$$= \frac{2}{2} \cdot \frac{3x+4}{x-2} - \frac{x-3}{2(x-2)}$$

$$= \frac{2(3x+4)-(x-3)}{2(x-2)}$$

$$= \frac{6x+8-x+3}{2(x-2)}$$

$$= \frac{5x+11}{2(x-2)}$$

Extra Practice

1. Find the LCD. Do not combine fractions.

$$\frac{7}{2x^2-11x+12}, \; \frac{13}{2x^2+7x-15}$$

2. Perform the operation indicated. Be sure to simplify. $\dfrac{5x}{x+1} - \dfrac{2x-5}{x+1}$

3. Perform the operation indicated. Be sure to simplify. $\dfrac{7}{x^2-y^2} + \dfrac{2y}{xy^2+y^3}$

4. Perform the operation indicated. Be sure to simplify. $\dfrac{1}{x^2+3x+2} - \dfrac{2}{x^2-x-6}$

Concept Check

Explain how to find the LCD of the fractions $\dfrac{3}{10xy^2z}$ and $\dfrac{6}{25x^2yz^3}$.

Chapter 7 Rational Expressions and Equations
7.4 Simplifying Complex Rational Expressions

Vocabulary

complex rational expression • complex fraction • numerator • denominator • LCD

1. A complex rational expression is also called a(n) _____.

2. A(n) _____ has a fraction in the numerator or in the denominator, or both.

3. A complex rational expression may contain two or more fractions in the _____ and denominator.

4. To simplify complex rational expressions, multiply the numerator and denominator of the complex fraction by the _____ of all the denominators appearing in the complex fraction.

Example	**Student Practice**
1. Simplify. $\dfrac{\dfrac{1}{x}}{\dfrac{2}{y^2}+\dfrac{1}{y}}$	2. Simplify. $\dfrac{\dfrac{1}{n^2}}{\dfrac{1}{m}+\dfrac{3}{4}}$

Add the two fractions in the denominator.

$$\frac{\dfrac{1}{x}}{\dfrac{2}{y^2}+\dfrac{1}{y}\cdot\dfrac{y}{y}}=\frac{\dfrac{1}{x}}{\dfrac{2+y}{y^2}}$$

Divide the fraction in the numerator by the fraction in the denominator.

$$\frac{1}{x}\div\frac{2+y}{y^2}=\frac{1}{x}\cdot\frac{y^2}{2+y}$$

$$=\frac{y^2}{x(2+y)}$$

Vocabulary Answers: 1. complex fraction 2. complex rational expression 3. numerator 4. LCD

Example	Student Practice
3. Simplify. $\dfrac{\dfrac{1}{x}+\dfrac{1}{y}}{\dfrac{3}{a}-\dfrac{2}{b}}$	**4.** Simplify. $\dfrac{\dfrac{1}{m}+\dfrac{1}{n}}{\dfrac{x}{5}-\dfrac{y}{3}}$

Observe that the LCD of the fractions in the numerator is xy. The LCD of the fractions in the denominator is ab.

$$\frac{\dfrac{1}{x}+\dfrac{1}{y}}{\dfrac{3}{a}-\dfrac{2}{b}}=\frac{\dfrac{1}{x}\cdot\dfrac{y}{y}+\dfrac{1}{y}\cdot\dfrac{x}{x}}{\dfrac{3}{a}\cdot\dfrac{b}{b}-\dfrac{2}{b}\cdot\dfrac{a}{a}}=\frac{\dfrac{y+x}{xy}}{\dfrac{3b-2a}{ab}}$$

$$=\frac{y+x}{xy}\cdot\frac{ab}{3b-2a}=\frac{ab(y+x)}{xy(3b-2a)}$$

| **5.** Simplify. $\dfrac{\dfrac{1}{x^2-1}+\dfrac{2}{x+1}}{x}$ | **6.** Simplify. $\dfrac{\dfrac{3x}{x^2+8x+12}-\dfrac{2}{x+6}}{5}$ |

We need to factor x^2-1.

$$\frac{\dfrac{1}{x^2-1}+\dfrac{2}{x+1}}{x}$$

$$=\frac{\dfrac{1}{(x+1)(x-1)}+\dfrac{2}{x+1}\cdot\dfrac{x-1}{x-1}}{x}$$

$$=\frac{\dfrac{1+2x-2}{(x+1)(x-1)}}{x}$$

$$=\frac{2x-1}{(x+1)(x-1)}\cdot\frac{1}{x}$$

$$=\frac{2x-1}{x(x+1)(x-1)}$$

Example	Student Practice

7. Simplify. $\dfrac{\dfrac{3}{a+b}-\dfrac{3}{a-b}}{\dfrac{5}{a^2-b^2}}$

$$=\frac{\dfrac{3}{a+b}\cdot\dfrac{a-b}{a-b}-\dfrac{3}{a-b}\cdot\dfrac{a+b}{a+b}}{\dfrac{5}{a^2-b^2}}$$

$$=\frac{\dfrac{3a-3b}{(a+b)(a-b)}-\dfrac{3a+3b}{(a+b)(a-b)}}{\dfrac{5}{a^2-b^2}}$$

$$=\frac{\dfrac{-6b}{(a+b)(a-b)}}{\dfrac{5}{(a+b)(a-b)}}$$

$$=\frac{-6b}{(a+b)(a-b)}\cdot\frac{(a+b)(a-b)}{5}$$

$$=\frac{-6b}{5}\ \text{ or }\ -\frac{6b}{5}$$

8. Simplify. $\dfrac{\dfrac{5}{x+y}+\dfrac{2}{x-y}}{\dfrac{7}{x^2-y^2}}$

9. Simplify by multiplying by the LCD.

$$\frac{\dfrac{5}{ab^2}-\dfrac{2}{ab}}{3-\dfrac{5}{2a^2b}}$$

$$=\frac{2a^2b^2\left(\dfrac{5}{ab^2}-\dfrac{2}{ab}\right)}{2a^2b^2\left(3-\dfrac{5}{2a^2b}\right)}$$

$$=\frac{2a^2b^2\left(\dfrac{5}{ab^2}\right)-2a^2b^2\left(\dfrac{2}{ab}\right)}{2a^2b^2(3)-2a^2b^2\left(\dfrac{5}{2a^2b}\right)}$$

$$=\frac{10a-4ab}{6a^2b^2-5b}$$

10. Simplify by multiplying by the LCD.

$$\frac{\dfrac{4}{y}-\dfrac{7}{4x}}{5-\dfrac{9}{xy^2}}$$

Example	Student Practice
11. Simplify by multiplying by the LCD. $$\dfrac{\dfrac{3}{a+b}-\dfrac{3}{a-b}}{\dfrac{5}{a^2-b^2}}$$	**12.** Simplify by multiplying by the LCD. $$\dfrac{\dfrac{5}{x+y}+\dfrac{2}{x-y}}{\dfrac{7}{x^2-y^2}}$$

The LCD of all individual fractions in the complex fraction is $(a+b)(a-b)$.

$$=\dfrac{(a+b)(a-b)\left(\dfrac{3}{a+b}\right)-(a+b)(a-b)\left(\dfrac{3}{a-b}\right)}{(a+b)(a-b)\left(\dfrac{5}{(a+b)(a-b)}\right)}$$

$$=\dfrac{3(a-b)-3(a+b)}{5}$$

$$=\dfrac{3a-3b-3a-3b}{5}$$

$$=-\dfrac{6b}{5}$$

Extra Practice

1. Simplify. $\dfrac{\dfrac{4}{a}+\dfrac{5}{b}}{\dfrac{2}{ab}}$

2. Simplify. $\dfrac{\dfrac{2}{x}+\dfrac{2}{y}}{x+y}$

3. Simplify. $\dfrac{\dfrac{2x}{x^2-25}}{\dfrac{4}{x+5}-\dfrac{3}{x-5}}$

4. Simplify. $\dfrac{\dfrac{3}{4x}-\dfrac{8}{3y}}{\dfrac{7}{a}+\dfrac{5}{b}}$

Concept Check

To simplify the following complex fraction, explain how you would add the two fractions in the numerator.

$$\dfrac{\dfrac{7}{x-3}+\dfrac{15}{2x-6}}{\dfrac{2}{x+5}}$$

Chapter 7 Rational Expressions and Equations
7.5 Solving Equations Involving Rational Expressions

Vocabulary

extraneous solution   •   no solution   •   LCD   •   exclude

1. The first step to solve an equation containing rational expressions is to determine the _____ of all the denominators.

2. If all of the apparent solutions of an equation are extraneous solutions, we say that the equation has _____.

3. A value that makes a denominator in the equation equal to zero is called a(n) _____.

4. _____ from your solution any value that would make the LCD equal to zero.

Example	**Student Practice**
1. Solve for x and check your solution. $$\frac{5}{x}+\frac{2}{3}=-\frac{3}{x}$$ The LCD is $3x$. Multiply each term by $3x$. $$3x\left(\frac{5}{x}\right)+3x\left(\frac{2}{3}\right)=3x\left(-\frac{3}{x}\right)$$ $$15+2x=-9$$ $$2x=-9-15$$ $$2x=-24$$ $$x=-12$$ Check. Replace each x by -12. $$\frac{5}{-12}+\frac{2}{3}\overset{?}{=}-\frac{3}{-12}$$ $$-\frac{5}{12}+\frac{8}{23}\overset{?}{=}\frac{3}{12}$$ $$\frac{3}{12}=\frac{3}{12}$$	**2.** Solve for x and check your solution. $$\frac{5}{x}+\frac{1}{4}=-\frac{6}{x}$$

Vocabulary Answers: 1. LCD 2. no solution 3. extraneous solution 4. exclude

Example	Student Practice

3. Solve and check. $\dfrac{6}{x+3} = \dfrac{3}{x}$

4. Solve and check. $\dfrac{2}{x+4} = \dfrac{6}{x}$

Observe that the LCD $= x(x+3)$.

$$x(x+3)\left(\dfrac{6}{x+3}\right) = x(x+3)\left(\dfrac{3}{x}\right)$$

$$6x = 3(x+3)$$

$$6x = 3x+9$$

$$3x = 9$$

$$x = 3$$

Check. Replace each x by 3.

$$\dfrac{6}{3+3} \overset{?}{=} \dfrac{3}{3}$$

$$\dfrac{6}{6} = \dfrac{3}{3}$$

---

5. Solve and check. $\dfrac{3}{x+5} - 1 = \dfrac{4-x}{2x+10}$

6. Solve and check.

$$\dfrac{3}{x+2} - \dfrac{5}{x+5} = \dfrac{x-2}{x^2+7x+10}$$

$$\dfrac{3}{x+5} - 1 = \dfrac{4-x}{2(x+5)}$$

$$2(x+5)\left(\dfrac{3}{x+5}\right) - 2(x+5)(1)$$

$$= 2(x+5)\left[\dfrac{4-x}{2(x+5)}\right]$$

$$2(3) - 2(x+5) = 4-x$$

$$6 - 2x - 10 = 4-x$$

$$-2x - 4 = 4-x$$

$$-4 = 4+x$$

$$-8 = x$$

The check is left to the student.

Example	Student Practice

7. Solve and check. $\dfrac{y}{y-2} - 4 = \dfrac{2}{y-2}$

8. Solve and check. $\dfrac{3x}{x-3} + 5 = \dfrac{3x}{x-3}$

Observe that the LCD is $y-2$.

$$(y-2)\left(\dfrac{y}{y-2}\right) - (y-2)(4)$$

$$= (y-2)\left(\dfrac{2}{y-2}\right)$$

$$y - 4(y-2) = 2$$

$$y - 4y + 8 = 2$$

$$-3y + 8 = 2$$

$$-3y = -6$$

$$\dfrac{-3y}{-3} = \dfrac{-6}{-3}$$

$$y = 2$$

This equation has no solution. We can see immediately that $y = 2$ is not a solution of the original equation. When we substitute 2 for y in a denominator, the denominator is equal to zero and the expression is undefined. Suppose that we tried to check the apparent solution by substituting 2 for y.

$$\dfrac{y}{y-2} - 4 = \dfrac{2}{y-2}$$

$$\dfrac{2}{2-2} - 4 \overset{?}{=} \dfrac{2}{2-2}$$

$$\dfrac{2}{0} - 4 = \dfrac{2}{0}$$

These expressions are not defined. There is no such number as $2 \div 0$. We see that 2 does not check. This equation has no solution.

Extra Practice

1. Solve and check. If there is no solution, so indicate. $\dfrac{2}{3x} + \dfrac{1}{6} = \dfrac{6}{x}$

2. Solve and check. If there is no solution, so indicate. $\dfrac{4}{a^2 - 1} = \dfrac{2}{a+1} + \dfrac{2}{a-1}$

3. Solve and check. If there is no solution, so indicate.
$$\dfrac{6}{5a + 10} - \dfrac{1}{a - 5} = \dfrac{4}{a^2 - 3a - 10}$$

4. Solve and check. If there is no solution, so indicate.
$$\dfrac{x-2}{x^2 - 4x - 5} + \dfrac{x+5}{x^2 - 25} = \dfrac{2x + 13}{x^2 + 6x + 5}$$

Concept Check

Explain how to find the LCD for the following equation. Do not solve the equation.
$$\dfrac{x}{x^2 - 9} + \dfrac{2}{3x - 9} = \dfrac{5}{2x + 6} + \dfrac{3}{2x^2 - 18}$$

Chapter 7 Rational Expressions and Equations
7.6 Ratio, Proportion, and Other Applied Problems

Vocabulary

ratio   •   proportion   •   cross multiplying   •   similar

1. A _____ is an equation that states that two ratios are equal.

2. _____ triangles are triangles that have the same shape, but may be different sizes.

3. If $\dfrac{a}{b} = \dfrac{c}{d}$, then $ad = bc$ is sometimes called _____.

4. A _____ is a comparison of two quantities.

Example	Student Practice
1. Michael took 5 hours to drive 245 miles on the turnpike. At the same rate, how many hours will it take him to drive a distance of 392 miles? Let $x =$ the number of hours it will take to drive 392 miles. If 5 hours are needed to drive 245 miles, then x hours are needed to drive 392 miles. $$\frac{5 \text{ hours}}{245 \text{ miles}} = \frac{x \text{ hours}}{392 \text{ miles}}$$ $$5(392) = 245x$$ $$\frac{1960}{245} = x$$ $$8 = x$$ It will take Michael 8 hours to drive 392 miles. To check, do the computations and see if $\dfrac{5}{245} = \dfrac{8}{392}$.	**2.** Rose took 4 hours to drive 264 miles. How far will she drive in 9 hours?

Vocabulary Answers: 1. proportion 2. similar 3. cross multiplying 4. ratio

Example	Student Practice

3. A ramp is 32 meters long and rises up 15 meters. A ramp at the same angle is 9 meters long. How high does the second ramp rise?

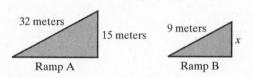

Ramp A   Ramp B

$$\frac{32}{9} = \frac{15}{x}$$

$$32x = (9)(15)$$

$$32x = 135$$

$$x = \frac{135}{32} \text{ or } x = 4\frac{7}{32}$$

4. Triangle M is similar to Triangle N. Find the length of side x. Express your answer as a mixed number.

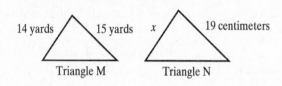

Triangle M   Triangle N

---

5. A French commuter airline flies from Paris to Avignon. Plane A flies at a speed that is 50 kilometers per hour faster than plane B. Plane A flies 500 kilometers in the amount of time that plane B flies 400 kilometers. Find the speed of each plane.

Let $s =$ the speed of plane B in kilometers per hour. Then $s + 50 =$ the speed of plane A in kilometers per hour. Each plane flies the same amount of time. That is, the time for plane A equals the time for plane B.

$$\frac{500}{s+50} = \frac{400}{s}$$

$$500s = (s+50)(400)$$

$$500s = 400s + 20,000$$

$$100s = 20,000$$

$$s = 200$$

Plane B travels 200 kilometers per hour. Since $s + 50 = 200 + 50 = 250,$ plane A travels 250 kilometers per hour.

6. Two trains travel in opposite directions for the same amount of time. Train A traveled 200 kilometers, while train B traveled 175 kilometers. Train A traveled 15 kilometers per hour faster than train B. What is the speed of each train?

Example	Student Practice
7. Reynaldo can sort a huge stack of mail on an old sorting machine in 9 hours. His brother Carlos can sort the same amount of mail using a newer sorting machine in 8 hours. How long will it take them to do the job working together? Express your answer in hours and minutes. Round to the nearest minute.	**8.** Two people are painting a side of a house. The first person takes 4 hours to paint the side of the house, while the second person takes 6 hours. How long will it take both of them to paint the side of the house working together? Express your answer in hours and minutes. Round to the nearest minute.

If Reynaldo can do the job in 9 hours, then in 1 hour he could do $\frac{1}{9}$ of the job. If Carlos can do the job in 8 hours, then in 1 hour he could do $\frac{1}{8}$ of the job. Let $x =$ the number of hours it takes Reynaldo and Carlos to do the job together. In 1 hour together they could do $\frac{1}{x}$ of the job. The amount of work Reynaldo can do in 1 hour plus the amount of work Carlos can do in 1 hour must be equal to the amount of work they could do together in 1 hour.

$$\frac{1}{9}+\frac{1}{8}=\frac{1}{x}$$

$$72x\left(\frac{1}{9}\right)+72x\left(\frac{1}{8}\right)=72x\left(\frac{1}{x}\right)$$

$$8x+9x=72$$

$$x=4\frac{4}{17}$$

To change $\frac{4}{17}$ of an hour to minutes,

$$\frac{4}{17}\ \text{hour} \times \frac{60\ \text{minutes}}{1\ \text{hour}} = \frac{240}{17}\ \text{minutes},$$

which is approximately 14.118. Thus doing the job together will take 4 hours and 14 minutes.

Extra Practice

1. Solve. $\dfrac{3}{10} = \dfrac{11.4}{x}$

2. The scale on a map of Massachusetts is approximately $\dfrac{3}{4}$ inch to 25 miles. If the distance from a college to Boston measures 4 inches on the map, how far apart are the two locations?

3. Alicia is 4 feet tall and casts a shadow that is 9 feet long. At the same time of day, a tree casts a shadow that is 36 feet long. How tall is the tree?

4. Toshiro and Mathilda work in the library. Working alone, it takes Toshiro 7 hours to arrange and stack 1000 books, while it takes Mathilda 6.5 hours. To the nearest minute, how long will the job take if they work together to arrange and stack 1000 books?

Concept Check

Mike found that his car used 18 gallons of gas to travel 396 miles. He needs to take a trip of 450 miles and wants to know how many gallons of gas it will take. He set up the equation $\dfrac{18}{x} = \dfrac{450}{396}$. Explain what error he made and how he should correctly solve the problem.

MATH COACH

Mastering the skills you need to do well on the test.

Watch the **MATH COACH** videos in MyMathLab® or on YouTube™ while you work the problems below. These helpful hints will help you avoid making common errors on test problems.

Dividing Rational Expressions—Problem 5 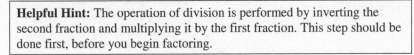 $\dfrac{2a^2-3a-2}{a^2+5a+6} \div \dfrac{a^2-5a+6}{a^2-9}$

> **Helpful Hint:** The operation of division is performed by inverting the second fraction and multiplying it by the first fraction. This step should be done first, before you begin factoring.

Did you keep the first fraction the same as it is written and then multiply it by $\dfrac{a^2-9}{a^2-5a+6}$? Yes ____ No ____

If you answered No, stop and make this correction to your work.

Were you able to factor each expression so that you obtained $\dfrac{(2a+1)(a-2)}{(a+2)(a+3)} \cdot \dfrac{(a+3)(a-3)}{(a-2)(a-3)}$? Yes ____ No ____

If you answered No, go back and check each factoring step to see if you can get the same result. Complete the problem by dividing out any common factors.

If you answered Problem 5 incorrectly, go back and rework the problem using these suggestions.

Subtracting Rational Expressions with Different Denominators—Problem 8 $\dfrac{3x}{x^2-3x-18} - \dfrac{x-4}{x-6}$

> **Helpful Hint:** First factor the trinomial in the denominator so that you can determine the LCD of these two fractions. Then multiply by what is needed in the second fraction so that it becomes an equivalent fraction with the LCD as the denominator. When subtracting, it is a good idea to place brackets around the second numerator to avoid sign errors.

Did you factor the first denominator into $(x-6)(x+3)$?
Yes ____ No ____

Did you then determine that the LCD is $(x-6)(x+3)$?
Yes ____ No ____

If you answered No to these questions, stop and review how to factor the trinomial in the first denominator and how to find the LCD when working with polynomials as denominators.

Did you multiply the numerator and denominator of the second fraction by $(x+3)$ and place brackets around the

product to obtain $\dfrac{3x}{(x-6)(x+3)} - \dfrac{[(x-4)(x+3)]}{(x-6)(x+3)}$?

Yes ____ No ____

Did you obtain $3x-x^2-x-12$ in the numerator? Yes ____ No ____

If you answered Yes, you forgot to distribute the negative sign. Rework the problem and make this correction.

As your final step, remember to combine like terms and then factor the new numerator before dividing out common factors.

Now go back and rework the problem using these suggestions.

Simplifying Complex Rational Expressions—Problem 10 Simplify. $\dfrac{\dfrac{6}{b}-4}{\dfrac{5}{bx}-\dfrac{10}{3x}}$

> **Helpful Hint:** There are two ways to simplify this expression:
> 1) combine the numerators and the denominators separately, or
> 2) multiply the numerator and denominator of the complex fraction by the LCD of all the denominators.
> Consider both methods and choose the one that seems easiest to you. We will show the steps of the first method for this particular problem.

Did you multiply 4 by $\dfrac{b}{b}$ and then subtract $\dfrac{6}{b}-\dfrac{4b}{b}$?

Yes _____   No _____

If you answered No, remember that 4 can be written as $\dfrac{4}{1}$, and to find the LCD, you must multiply numerator and denominator by the variable b.

In the denominator of the complex fraction, did you obtain $3bx$ as the LCD and multiply the first fraction by $\dfrac{3}{3}$ and the second fraction by $\dfrac{b}{b}$ before subtracting the two fractions?

Yes _____   No _____

If you answered No, stop and carefully change these two fractions into equivalent fractions with $3bx$ as the

denominator. Then subtract the numerators and keep the common denominator.

Were you able to rewrite the problem as follows: $\dfrac{6-4b}{b}\div\dfrac{15-10b}{3bx}$?

Yes _____   No _____

If you answered No, examine your steps carefully. Once you have one fraction in the numerator and one fraction in the denominator, you can rewrite the division as multiplication by inverting the fraction in the denominator.

If you answered Problem 10 incorrectly, go back and rework the problem using these suggestions.

Solving Equations Involving Rational Expressions—Problem 14

Solve for x. $\dfrac{x-3}{x-2}=\dfrac{2x^2-15}{x^2+x-6}-\dfrac{x+1}{x+3}$

> **Helpful Hint:** First factor any denominators that need to be factored so that you can determine the LCD of all denominators. Verify that you are solving an equation, then multiply each term of the equation by the LCD. Solve the resulting equation. Check your solution.

Did you factor x^2+x-6 to get $(x+3)(x-2)$ and identify the LCD as $(x+3)(x-2)$? Yes _____   No _____
If you answered No, go back and complete these steps again.

Did you notice that we are solving an equation and that we can multiply each term of the equation by the LCD and then multiply the binomials to obtain the equation

$x^2-9=(2x^2-15)-(x^2-x-2)$? Yes _____   No _____

If you answered No, look at the step $-\left[(x+1)(x-2)\right]$.

After you multiplied the binomials, did you distribute the negative sign and change the sign of all terms inside the grouping symbols?
Yes _____   No _____

If you answered No, review how to multiply binomials and subtract polynomial expressions. Then combine like terms and solve the equation for x.

Now go back and rework the problem using these suggestions.

Chapter 8 Rational Exponents
8.1 Simplifying Expressions with Rational Exponents

Vocabulary

rational exponents • denominator • quotient rule • exponential factor

1. You need to change fractional exponents to equivalent fractional exponents with the same _____ when the rules of exponents require you to add or subtract them.

2. Exponents that are fractions are called _____.

3. When factoring an expression, if the terms contain exponents, we look for the same _____ in each term.

Example	Student Practice
1. Simplify. $\left(\dfrac{5xy^{-3}}{2x^{-4}y}\right)^{-2}$	**2.** Simplify. $\left(\dfrac{4a^{-3}b}{3a^{-2}b^{-5}}\right)^{-3}$

Apply the rules of exponents.

$$\left(\frac{5xy^{-3}}{2x^{-4}y}\right)^{-2} = \frac{\left(5xy^{-3}\right)^{-2}}{\left(2x^{-4}y\right)^{-2}} \qquad \left(\frac{x}{y}\right)^{n} = \frac{x^{n}}{y^{n}}$$

$$= \frac{5^{-2}x^{-2}\left(y^{-3}\right)^{-2}}{2^{-2}\left(x^{-4}\right)^{-2}y^{-2}} \qquad (xy)^{n} = x^{n}y^{n}$$

$$= \frac{5^{-2}x^{-2}y^{6}}{2^{-2}x^{8}y^{-2}} \qquad \left(x^{m}\right)^{n} = x^{mn}$$

$$= \frac{5^{-2}}{2^{-2}} \cdot \frac{x^{-2}}{x^{8}} \cdot \frac{y^{6}}{y^{-2}}$$

$$= \frac{2^{2}}{5^{2}} \cdot \frac{x^{-2}}{x^{8}} \cdot \frac{y^{6}}{y^{-2}} \qquad \frac{x^{-n}}{y^{-m}} = \frac{y^{m}}{x^{n}}$$

$$= \frac{2^{2}}{5^{2}} \cdot x^{-2-8} \cdot y^{6+2} \qquad \frac{x^{m}}{x^{n}} = x^{m-n}$$

$$= \frac{4}{25}x^{-10}y^{8}$$

Vocabulary Answers: 1. denominator 2. rational exponents 3. exponential factor

Example	Student Practice
3. Simplify.	**4.** Simplify.
(a) $\left(x^{2/3}\right)^4$	**(a)** $\left(x^3\right)^{4/5}$
$\left(x^{2/3}\right)^4 = x^{(2/3)(4/1)} = x^{8/3}$	
(b) $\dfrac{x^{5/6}}{x^{1/6}}$	**(b)** $\dfrac{x^{7/8}}{x^{1/8}}$
$\dfrac{x^{5/6}}{x^{1/6}} = x^{5/6-1/6} = x^{4/6} = x^{2/3}$	**(c)** $6^{5/13} \cdot 6^{4/13}$
(c) $x^{2/3} \cdot x^{-1/3}$	
$x^{2/3} \cdot x^{-1/3} = x^{2/3-1/3} = x^{1/3}$	

Example	Student Practice
5. Simplify. Express your answers with positive exponents only.	**6.** Simplify. Express your answers with positive exponents only.
(a) $\left(2x^{1/2}\right)\left(3x^{1/3}\right)$	**(a)** $\left(-3x^{2/5}\right)\left(4x^{1/2}\right)$
$\left(2x^{1/2}\right)\left(3x^{1/3}\right) = 6x^{1/2+1/3}$	
$= 6x^{3/6+2/6} = 6x^{5/6}$	
(b) $\dfrac{18x^{1/4}y^{-1/3}}{-6x^{-1/2}y^{1/6}}$	**(b)** $\dfrac{25x^{3/4}y^{-5/12}}{5x^{-1/8}y^{1/4}}$
$\dfrac{18x^{1/4}y^{-1/3}}{-6x^{-1/2}y^{1/6}} = -3x^{1/4-(-1/2)}y^{-1/3-1/6}$	
$= -3x^{1/4+2/4}y^{-2/6-1/6}$	
$= -3x^{3/4}y^{-3/6}$	
$= -3x^{3/4}y^{-1/2}$	
$= -\dfrac{3x^{3/4}}{y^{1/2}}$	

Example	Student Practice
7. Multiply and simplify. $-2x^{5/6}\left(3x^{1/2}-4x^{-1/3}\right)$ We will need to be very careful when we add the exponents for x as we use the distributive property. $-2x^{5/6}\left(3x^{1/2}-4x^{-1/3}\right)$ $=-6x^{5/6+1/2}+8x^{5/6-1/3}$ $=-6x^{5/6+3/6}+8x^{5/6-2/6}$ $=-6x^{8/6}+8x^{3/6}$ $=-6x^{4/3}+8x^{1/2}$	**8.** Multiply and simplify. $4x^{2/3}\left(-2x^{1/2}+6x^{-1/6}\right)$
9. Evaluate. **(a)** $\left(25\right)^{3/2}$ $\left(25\right)^{3/2}=\left(5^2\right)^{3/2}=5^{2/1\cdot3/2}=5^3=125$ **(b)** $\left(27\right)^{2/3}$ $\left(27\right)^{2/3}=\left(3^3\right)^{2/3}=3^{3/1\cdot2/3}=3^2=9$	**10.** Evaluate. **(a)** $\left(16\right)^{3/2}$ **(b)** $\left(64\right)^{2/3}$
11. Write as one fraction with positive exponents. $2x^{-1/2}+x^{1/2}$ $2x^{-1/2}+x^{1/2}=\dfrac{2}{x^{1/2}}+\dfrac{x^{1/2}\cdot x^{1/2}}{x^{1/2}}$ $=\dfrac{2}{x^{1/2}}+\dfrac{x^1}{x^{1/2}}$ $=\dfrac{2+x}{x^{1/2}}$	**12.** Write as one fraction with positive exponents. $5x^{-3/4}+x^{1/4}$

Example	Student Practice
13. Factor out the common factor $2x$. $2x^{3/2} + 4x^{5/2}$ Rewrite the exponent of each term so that each term contains the factor $2x$ or $2x^{2/2}$. $2x^{3/2} + 4x^{5/2}$ $= 2x^{2/2+1/2} + 4x^{2/2+3/2}$ $= 2\left(x^{2/2}\right)\left(x^{1/2}\right) + 4\left(x^{2/2}\right)\left(x^{3/2}\right)$ $= 2x\left(x^{1/2} + 2x^{3/2}\right)$	**14.** Factor out the common factor $6z$. $18z^{5/3} - 24z^{4/3}$

Extra Practice

1. Simplify. Express your answers with positive exponents only.

$$\left(x^{4/5}\right)^4$$

2. Simplify. Express your answers with positive exponents only.

$$\left(a^{5/4}b^{2/3}\right)\left(a^{-1/4}b^{2/3}\right)$$

3. Write as one fraction with positive exponents.

$$y^{-2/3} + 3y^{1/3}$$

4. Factor out the common factor $5x$.

$$10x^{7/4} + 25x^{9/8}$$

Concept Check

Explain how you would simplify the following. $x^{9/10} \cdot x^{-1/5}$

Chapter 8 Rational Exponents
8.2 Radical Expressions and Functions

Vocabulary
square root • principal square root • radical sign • radical expression • radicand
index • higher-order roots • cube root • fourth root • square root function

1. Because the symbol $\sqrt{x}$ represents exactly one real number for all real numbers x that are nonnegative, we can use it to define the _____, $f(x) = \sqrt{x}$.

2. The positive square root is the _____.

3. A(n) _____ of a number is a value that when cubed is equal to the original number.

4. In the radical expression $\sqrt[n]{x}$, x is the radicand and n is the _____.

Example	Student Practice
1. Find the indicated function values of the function $f(x) = \sqrt{2x+4}$. Round your answers to the nearest tenth when necessary.	**2.** Find the indicated function values of the function $f(x) = \sqrt{5x-4}$. Round your answers to the nearest tenth when necessary.
(a) $f(-2)$ Substitute -2 for x in the function. $f(-2) = \sqrt{2(-2)+4} = \sqrt{-4+4}$ $= \sqrt{0} = 0$	**(a)** $f(4)$
(b) $f(6)$ $f(6) = \sqrt{2(6)+4} = \sqrt{12+4}$ $= \sqrt{16} = 4$	**(b)** $f(1)$
(c) $f(3)$ $f(3) = \sqrt{2(3)+4} = \sqrt{6+4}$ $= \sqrt{10} \approx 3.2$	**(c)** $f(3)$

Vocabulary Answers: 1. square root function 2. principal square root 3. cube root 4. index

Example	Student Practice
3. Find the domain of the function. $f(x) = \sqrt{3x - 6}$ The expression $3x - 6$ must be nonnegative. $3x - 6 \geq 0$ $\qquad 3x \geq 6$ $\qquad x \geq 2$ Thus, the domain is all real numbers x, where $x \geq 2$.	**4.** Find the domain of the function. $f(x) = \sqrt{\dfrac{2}{3}x - 6}$
5. Change $\sqrt[4]{x^4}$ to rational exponents and simplify. Assume that all variables are nonnegative real numbers. If n is a positive integer and x is a nonnegative real number, then $x^{1/n} = \sqrt[n]{x}$. $\sqrt[4]{x^4} = \left(x^4\right)^{1/4} = x^{4/4} = x^1 = x$	**6.** Change $\sqrt[6]{7^6}$ to rational exponents and simplify. Assume that all variables are nonnegative real numbers.
7. Replace all radicals with rational exponents. **(a)** $\sqrt[3]{x^2}$ $\qquad \sqrt[3]{x^2} = \left(x^2\right)^{1/3}$ $\qquad\qquad = x^{2/3}$ **(b)** $\left(\sqrt[5]{w}\right)^7$ $\qquad \left(\sqrt[5]{w}\right)^7 = \left(w^{1/5}\right)^7$ $\qquad\qquad = w^{7/5}$	**8.** Replace all radicals with rational exponents. **(a)** $\sqrt[5]{z^3}$ **(b)** $\left(\sqrt[3]{a}\right)^8$

Example	Student Practice
9. Change to radical form.	**10.** Change to radical form.
(a) $w^{-2/3}$	**(a)** $(ab)^{-5/3}$

For positive integers m and n and any real number x where $x \neq 0$ and for which $x^{1/n}$ is defined, then

$$x^{-m/n} = \frac{1}{x^{m/n}} = \frac{1}{\left(\sqrt[n]{x}\right)^m} = \frac{1}{\sqrt[n]{x^m}}.$$

$$w^{-2/3} = \frac{1}{w^{2/3}} = \frac{1}{\left(\sqrt[3]{w}\right)^2} = \frac{1}{\sqrt[3]{w^2}}$$

| **(b)** $(3x)^{3/4}$ | **(b)** $(5x)^{1/5}$ |

For positive integers m and n and any real number x for which $x^{1/n}$ is defined, then $x^{m/n} = \left(\sqrt[n]{x}\right)^m = \sqrt[n]{x^m}$.

$$(3x)^{3/4} = \sqrt[4]{(3x)^3} = \sqrt[4]{27x^3} \text{ or}$$

$$(3x)^{3/4} = \left(\sqrt[4]{3x}\right)^3$$

11. Change to radical form and evaluate.	**12.** Change to radical form and evaluate.
(a) $(-16)^{5/2}$	**(a)** $64^{-1/3}$

$(-16)^{5/2} = \left(\sqrt{-16}\right)^5$; however, $\sqrt{-16}$ is not a real number. Thus, $(-16)^{5/2}$ is not a real number.

| | **(b)** $(-81)^{1/4}$ |

(b) $144^{-1/2}$

$$144^{-1/2} = \frac{1}{144^{1/2}} = \frac{1}{\sqrt{144}} = \frac{1}{12}$$

Example	Student Practice
13. Simplify. Assume that x and y may be any real numbers.	**14.** Simplify. Assume that x and y may be any real numbers.
(a) $\sqrt{49x^2}$	**(a)** $\sqrt{25a^2}$
Since the index is even, take the absolute value. $$\sqrt{49x^2} = 7|x|$$	
(b) $\sqrt[4]{81y^{16}}$	**(b)** $\sqrt[3]{64z^6}$
$$\sqrt[4]{81y^{16}} = 3\left|y^4\right|$$ Since any number raised to the fourth is positive we can write it as $$\sqrt[4]{81y^{16}} = 3y^4.$$	
	(c) $\sqrt[4]{81x^8y^4}$
(c) $\sqrt[3]{27x^6y^9}$	
Note that the index is odd, thus we do not need the absolute value. $$\sqrt[3]{27x^6y^9} = \sqrt[3]{3^3\left(x^2\right)^3\left(y^3\right)^3} = 3x^2y^3$$	

Extra Practice

1. Evaluate if possible. $\sqrt[15]{(7)^{15}}$

2. Replace all radicals with rational exponents. Assume that variables represent positive real numbers. $$\sqrt[8]{(4m-3n)^5}$$

3. Change to radical form. $4^{-4/7}$

4. Simplify. $\sqrt{49a^6b^{16}}$

Concept Check

Explain how you would simplify the following. $\sqrt[4]{81x^8y^{16}}$

Name: _____   Date: _____

Instructor: _____   Section: _____

Chapter 8 Rational Exponents
8.3 Simplifying, Adding, and Subtracting Radicals

Vocabulary

product rule • radicand • index • like radicals • distributive property

1. Like radicals have the same _____ and index.

2. The_____ for radicals states that for all nonnegative real numbers a and b and positive integers n, $\sqrt[n]{a}\sqrt[n]{b} = \sqrt[n]{ab}$.

3. Only _____ can be added or subtracted.

Example	Student Practice
1. Simplify. $\sqrt{32}$ Recall that for all nonnegative real numbers a and b and positive integers n, $\sqrt[n]{a}\sqrt[n]{b} = \sqrt[n]{ab}$. Factor the radicand into the largest perfect square factors possible, then simplify. $\sqrt{32} = \sqrt{16 \cdot 2} = \sqrt{16}\sqrt{2} = 4\sqrt{2}$	**2.** Simplify. $\sqrt{50}$
3. Simplify. $\sqrt{48}$ $\sqrt{48} = \sqrt{16 \cdot 3} = \sqrt{16}\sqrt{3} = 4\sqrt{3}$	**4.** Simplify. $\sqrt{108}$
5. Simplify. $\sqrt[3]{-81}$ Factor the radicand into the largest perfect cube factors possible, then simplify. $\sqrt[3]{-81} = \sqrt[3]{-27}\sqrt[3]{3} = -3\sqrt[3]{3}$	**6.** Simplify. $\sqrt[3]{-162}$

Vocabulary Answers: 1. radicand 2. product rule 3. like radicals

Example	Student Practice
7. Simplify.	**8.** Simplify.

(a) $\sqrt{27x^3y^4}$

Factor out the perfect squares.

$$\sqrt{27x^3y^4} = \sqrt{9 \cdot 3 \cdot x^2 \cdot x \cdot y^4}$$
$$= \sqrt{9x^2y^4}\sqrt{3x} = 3xy^2\sqrt{3x}$$

(a) $\sqrt{32a^5b^4}$

(b) $\sqrt[3]{16x^4y^3z^6}$

Factor out the perfect cubes.

$$\sqrt[3]{16x^4y^3z^6} = \sqrt[3]{8 \cdot 2 \cdot x^3 \cdot x \cdot y^3 \cdot z^6}$$
$$= \sqrt[3]{8x^3y^3z^6}\sqrt[3]{2x}$$
$$= 2xyz^2\sqrt[3]{2x}$$

(b) $\sqrt[3]{192x^2y^4z^5}$

---

9. Combine. $2\sqrt{5} + 3\sqrt{5} - 4\sqrt{5}$

All three radicals are like radicals because they have the same radicand and index. Thus, they may be combined.

$$2\sqrt{5} + 3\sqrt{5} - 4\sqrt{5} = (2+3-4)\sqrt{5}$$
$$= 1\sqrt{5} = \sqrt{5}$$

10. Combine. $6\sqrt{3a} - 3\sqrt{3a} + \sqrt{3a}$.

---

11. Combine. $5\sqrt{3} - \sqrt{27} + 2\sqrt{48}$

Sometimes when simplifying radicands, you may find like radicals.

$$5\sqrt{3} - \sqrt{27} + 2\sqrt{48}$$
$$= 5\sqrt{3} - \sqrt{9}\sqrt{3} + 2\sqrt{16}\sqrt{3}$$
$$= 5\sqrt{3} - 3\sqrt{3} + 2(4)\sqrt{3}$$
$$= 5\sqrt{3} - 3\sqrt{3} + 8\sqrt{3}$$
$$= 10\sqrt{3}$$

12. Combine. $6\sqrt{5} + \sqrt{80} - 3\sqrt{20}$

Example	Student Practice
13. Combine. $6\sqrt{x} + 4\sqrt{12x} - \sqrt{75x} + 3\sqrt{x}$	**14.** Combine. $$3\sqrt{a} - 2\sqrt{32a} + \sqrt{50a} + 2\sqrt{a}$$

$6\sqrt{x} + 4\sqrt{12x} - \sqrt{75x} + 3\sqrt{x}$

$= 6\sqrt{x} + 4\sqrt{4}\sqrt{3x} - \sqrt{25}\sqrt{3x} + 3\sqrt{x}$

$= 6\sqrt{x} + 8\sqrt{3x} - 5\sqrt{3x} + 3\sqrt{x}$

$= 6\sqrt{x} + 3\sqrt{x} + 8\sqrt{3x} - 5\sqrt{3x}$

$= 9\sqrt{x} + 3\sqrt{3x}$

15. Combine. $2\sqrt[3]{81x^3 y^4} + 3xy\sqrt[3]{24y}$	**16.** Combine. $\sqrt[3]{40x^6 z^4} + 3x^2\sqrt[3]{320z^4}$

$2\sqrt[3]{81x^3 y^4} + 3xy\sqrt[3]{24y}$

$= 2\sqrt[3]{27x^3 y^3}\sqrt[3]{3y} + 3xy\sqrt[3]{8}\sqrt[3]{3y}$

$= 2(3xy)\sqrt[3]{3y} + 3xy(2)\sqrt[3]{3y}$

$= 6xy\sqrt[3]{3y} + 6xy\sqrt[3]{3y}$

$= 12xy\sqrt[3]{3y}$

Extra Practice

1. Simplify. Assume that all variables are nonnegative real numbers. $\sqrt{25x^3}$

2. Simplify. Assume that all variables are nonnegative real numbers. $\sqrt[4]{625ab^{19}}$

3. Combine. $6\sqrt{7} - 4\sqrt{5} + 4\sqrt{7}$

4. Combine. Assume that all variables represent nonnegative real numbers. $6\sqrt{48x^2} - 2\sqrt{27x^2} - \sqrt{3x^2}$

Concept Check

Explain how you would simplify the following. $\sqrt[4]{16x^{13} y^{16}}$

Chapter 8 Rational Exponents
8.4 Multiplying and Dividing Radicals

Vocabulary
product rule for radicals   •   FOIL method   •   quotient rule for radicals
rationalizing the denominator   •   conjugates

1. The expressions $a+b$ and $a-b$, where a and b represent any algebraic term, are called
   _____.

2. The _____ states that for all nonnegative real numbers a, all positive numbers
   b, and positive integers n, $\dfrac{\sqrt[n]{a}}{\sqrt[n]{b}} = \sqrt[n]{\dfrac{a}{b}}$.

3. _____ is the process of transforming a fraction with one or more radicals in the
   denominator into an equivalent fraction without a radical in the denominator.

Example	**Student Practice**
1. Multiply. $\left(3\sqrt{2}\right)\left(5\sqrt{11x}\right)$	**2.** Multiply. $\left(3\sqrt{7}\right)\left(-4\sqrt{3z}\right)$
To multiply, use the product rule for radicals, $\sqrt[n]{a}\,\sqrt[n]{b} = \sqrt[n]{ab}$.	
$\left(3\sqrt{2}\right)\left(5\sqrt{11x}\right) = (3)(5)\left(\sqrt{2\cdot 11x}\right)$ $= 15\sqrt{2\cdot 11x} = 15\sqrt{22x}$	
3. Multiply. $\left(\sqrt{2}+3\sqrt{5}\right)\left(2\sqrt{2}-\sqrt{5}\right)$	**4.** Multiply. $\left(4\sqrt{3}-3\sqrt{5}\right)\left(2\sqrt{3}-\sqrt{5}\right)$
To multiply two binomials containing radicals, we can use the distributive property or the FOIL method. For this problem, we use the FOIL method.	
$\left(\sqrt{2}+3\sqrt{5}\right)\left(2\sqrt{2}-\sqrt{5}\right)$ $= 2\sqrt{4} - \sqrt{10} + 6\sqrt{10} - 3\sqrt{25}$ $= 4 + 5\sqrt{10} - 15 = -11 + 5\sqrt{10}$	

Vocabulary Answers: 1. conjugates 2. quotient rule for radicals 3. rationalizing the denominator

Example	Student Practice
5. Multiply. $\left(7-3\sqrt{2}\right)\left(4-\sqrt{3}\right)$	**6.** Multiply. $\left(6+4\sqrt{3}\right)\left(2+3\sqrt{2}\right)$

Use the FOIL method.

$$\left(7-3\sqrt{2}\right)\left(4-\sqrt{3}\right)$$
$$= 28-7\sqrt{3}-12\sqrt{2}+3\sqrt{6}$$

---

7. Multiply. $\left(\sqrt{7}+\sqrt{3x}\right)^2$	**8.** Multiply. $\left(\sqrt{11}-\sqrt{2a}\right)^2$

Use $\left(a+b\right)^2 = a^2+2ab+b^2$, where $a = \sqrt{7}$ and $b = \sqrt{3x}$.

$$\left(\sqrt{7}+\sqrt{3x}\right)^2$$
$$= \left(\sqrt{7}\right)^2 + 2\sqrt{7}\sqrt{3x} + \left(\sqrt{3x}\right)^2$$
$$= 7+2\sqrt{21x}+3x$$

---

9. Divide.	**10.** Divide.
(a) $\sqrt[3]{\dfrac{125}{8}}$	**(a)** $\sqrt[3]{\dfrac{8}{27}}$

For all nonnegative real numbers a, all positive numbers b, and positive integers n, $\dfrac{\sqrt[n]{a}}{\sqrt[n]{b}} = \sqrt[n]{\dfrac{a}{b}}$.

$$\sqrt[3]{\frac{125}{8}} = \frac{\sqrt[3]{125}}{\sqrt[3]{8}} = \frac{5}{2}$$

(b) $\dfrac{\sqrt{28x^5y^3}}{\sqrt{7x}}$

(b) $\dfrac{\sqrt{72a^3b^7}}{\sqrt{2b^3}}$

$$\frac{\sqrt{28x^5y^3}}{\sqrt{7x}} = \sqrt{\frac{28x^5y^3}{7x}} = \sqrt{4x^4y^3}$$
$$= 2x^2y\sqrt{y}$$

Example	Student Practice
11. Simplify. $\dfrac{3}{\sqrt{12x}}$	**12.** Simplify. $\dfrac{5}{\sqrt{32z}}$

Simplify the radical in the denominator, then multiply in order to rationalize the denominator.

$$\frac{3}{\sqrt{12x}} = \frac{3}{\sqrt{4}\sqrt{3x}}$$

$$= \frac{3}{2\sqrt{3x}} \cdot \frac{\sqrt{3x}}{\sqrt{3x}}$$

$$= \frac{\cancel{3}\sqrt{3x}}{2(\cancel{3}x)} = \frac{\sqrt{3x}}{2x}$$

13. Simplify. $\sqrt[3]{\dfrac{2}{3x^2}}$	**14.** Simplify. $\sqrt[3]{\dfrac{54}{2a}}$

Multiply the numerator and denominator by a value that will make the radicand in the denominator a perfect cube (i.e., rationalize the denominator). In this case, multiply the numerator and denominator by $9x$ because $9x \cdot 3x^2 = 27x^3$ which is a perfect cube.

$$\sqrt[3]{\frac{2}{3x^2}} = \sqrt[3]{\frac{2}{3x^2} \cdot \frac{9x}{9x}}$$

$$= \sqrt[3]{\frac{18x}{27x^3}}$$

Rewrite with the quotient rule and simplify.

$$\sqrt[3]{\frac{18x}{27x^3}} = \frac{\sqrt[3]{18x}}{\sqrt[3]{27x^3}}$$

$$= \frac{\sqrt[3]{18x}}{3x}$$

Example	Student Practice
15. Simplify. $\dfrac{5}{3+\sqrt{2}}$	**16.** Simplify. $\dfrac{\sqrt{5}+3\sqrt{7}}{2\sqrt{5}-\sqrt{7}}$

Multiply the numerator and denominator by the conjugate of $3+\sqrt{2}$.

$$\frac{5}{3+\sqrt{2}} = \frac{5}{3+\sqrt{2}} \cdot \frac{3-\sqrt{2}}{3-\sqrt{2}}$$

$$= \frac{15-5\sqrt{2}}{3^2 - \left(\sqrt{2}\right)^2}$$

$$= \frac{15-5\sqrt{2}}{9-2} = \frac{15-5\sqrt{2}}{7}$$

Extra Practice

1. Multiply and simplify. Assume that all variables are nonnegative numbers.

$$\left(3a\sqrt{a}\right)\left(5\sqrt{b}\right)$$

2. Multiply and simplify. Assume that all variables are nonnegative numbers.

$$\left(5\sqrt{7}+3\sqrt{10}\right)^2$$

3. Divide and simplify. Assume that all variables represent positive numbers.

$$\sqrt{\frac{18x}{25y^4}}$$

4. Simplify by rationalizing the denominator. $\dfrac{2x}{\sqrt{5}-\sqrt{3}}$

Concept Check

Explain how you would rationalize the denominator of the following. $\dfrac{5\sqrt{3}-3\sqrt{2}}{3\sqrt{2}-2\sqrt{3}}$

Name: _____   Date: _____

Instructor: _____   Section: _____

Chapter 8 Rational Exponents
8.5 Radical Equations

Vocabulary
radical equation • isolating the radical term • zero factor • extraneous solution

1. If the result of squaring both sides of a radical equation is a quadratic equation, collect all terms on one side and use the _____ method to solve.

2. A(n) _____ is an equation with a variable in one or more of the radicals.

3. Getting one radical expression alone on one side of the equation is referred to as _____.

Example	Student Practice
1. Solve. $\sqrt{2x+9} = x+3$	**2.** Solve. $\sqrt{3x+7} = x-1$

1. Solve. $\sqrt{2x+9} = x+3$

Square each side and simplify.

$$\left(\sqrt{2x+9}\right)^2 = \left(x+3\right)^2$$
$$2x+9 = x^2 +6x+9$$

Collect all terms on one side and factor.
$$2x+9 = x^2 +6x+9$$
$$0 = x^2 +4x$$
$$0 = x\left(x+4\right)$$

Set each factor equal to zero and solve.
$$x = 0 \quad \text{or} \quad x+4=0$$
$$x = 0 \qquad\qquad x = -4$$

Check.

$$x=0: \ \sqrt{2(0)+9} \overset{?}{=} (0)+3$$
$$3 = 3$$
$$x=-4: \ \sqrt{2(-4)+9} \overset{?}{=} (-4)+3$$
$$1 \neq -1$$

Thus, 0 is the only solution.

Vocabulary Answers: 1. zero factor 2. radical equation 3. isolating the radical

Example	Student Practice

3. Solve. $\sqrt{10x+5}-1=2x$

Isolate the radical term. Then square each side and simplify.

$$\sqrt{10x+5}-1=2x$$
$$\sqrt{10x+5}=2x+1$$
$$\left(\sqrt{10x+5}\right)^2=(2x+1)^2$$
$$10x+5=4x^2+4x+1$$

Collect all terms on one side and factor.

$$10x+5=4x^2+4x+1$$
$$0=4x^2-6x-4$$
$$0=2\left(2x^2-3x-2\right)$$
$$0=2(2x+1)(x-2)$$

Set each factor equal to zero and solve.

$$2x+1=0 \qquad\qquad x-2=0$$
$$2x=-1 \qquad\qquad x=2$$
$$x=-\frac{1}{2}$$

Check.

$$x=-\frac{1}{2}: \quad \sqrt{10\left(-\frac{1}{2}\right)+5}-1\overset{?}{=}2\left(-\frac{1}{2}\right)$$
$$\sqrt{-5+5}-1\overset{?}{=}-1$$
$$\sqrt{0}-1\overset{?}{=}-1$$
$$-1=-1$$
$$x=2: \quad \sqrt{10(2)+5}-1\overset{?}{=}2(2)$$
$$\sqrt{25}-1\overset{?}{=}4$$
$$4=4$$

Both answers check. Thus, the solutions are $-\frac{1}{2}$ and 2.

4. Solve. $3x=-1+\sqrt{30x+10}$

Example	Student Practice

5. Solve. $\sqrt{5x+1} - \sqrt{3x} = 1$

Isolate one of the radicals.

$\sqrt{5x+1} - \sqrt{3x} = 1$
$\sqrt{5x+1} = 1 + \sqrt{3x}$

Square each side.

$\sqrt{5x+1} = 1 + \sqrt{3x}$
$\left(\sqrt{5x+1}\right)^2 = \left(1+\sqrt{3x}\right)^2$
$5x+1 = \left(1+\sqrt{3x}\right)\left(1+\sqrt{3x}\right)$
$5x+1 = 1 + 2\sqrt{3x} + 3x$

Isolate the remaining radical.

$5x+1 = 1 + 2\sqrt{3x} + 3x$
$2x = 2\sqrt{3x}$
$x = \sqrt{3x}$

Square each side.

$x = \sqrt{3x}$
$(x)^2 = \left(\sqrt{3x}\right)^2$
$x^2 = 3x$

Collect all terms on one side and factor.

$x^2 = 3x$
$x^2 - 3x = 0$
$x(x-3) = 0$
$x = 0$ or $x - 3 = 0$
$x = 0$ $x = 3$

The check is left to the student. Verify that both 0 and 3 are valid solutions.

6. Solve. $\sqrt{x+4} - 2 = \sqrt{x-4}$

Example	Student Practice
7. Solve. $\sqrt{2y+5}-\sqrt{y-1}=\sqrt{y+2}$	**8.** Solve. $\sqrt{y+6}-\sqrt{y-2}=\sqrt{2y-2}$

$$\sqrt{2y+5}-\sqrt{y-1}=\sqrt{y+2}$$
$$\left(\sqrt{2y+5}-\sqrt{y-1}\right)^2=\left(\sqrt{y+2}\right)^2$$

Expand both sides of the equation.
$$2y+5-2\sqrt{(y-1)(2y+5)}+y-1=y+2$$

Isolate the radical and solve for y.
$$-2\sqrt{(y-1)(2y+5)}=-2y-2$$
$$\sqrt{(y-1)(2y+5)}=y+1$$
$$\left(\sqrt{(y-1)(2y+5)}\right)^2=(y+1)^2$$
$$2y^2+3y-5=y^2+2y+1$$
$$y^2+y-6=0$$
$$(y+3)(y-2)=0$$
$$y=-3 \quad \text{or} \quad y=2$$

The check is left to the student. Verify that 2 is a valid solution but -3 is not.

Extra Practice

1. Solve the radical equation. Check your solution. $\sqrt{10x-5}=5$

2. Solve the radical equation. Check your solution. $\sqrt{x+4}+8=3$

3. Solve the radical equation. This will usually involve squaring each side twice. Check your solution. $\sqrt{3x+4}=2+\sqrt{x}$

4. Solve the radical equation. This will usually involve squaring each side twice. Check your solution.
$$\sqrt{2x+6}-\sqrt{x+4}=\sqrt{x-4}$$

Concept Check

When you try to solve the equation $2+\sqrt{x+10}=x$, you obtain the values $x=-1$ and $x=6$. Explain how you would determine if either of these values is a solution of the radical equation.

Name: _____   Date: _____

Instructor: _____   Section: _____

Chapter 8 Rational Exponents
8.6 Complex Numbers

Vocabulary
imaginary number • complex number • real part • imaginary part • conjugates

1. The complex numbers $a+bi$ and $a-bi$ are called _____.

2. The _____ i is defined as $i = \sqrt{-1}$ and $i^2 = -1$.

3. A number that can be written in the form $a+bi$, where a and b are real numbers, is a _____.

4. In the complex number $a+bi$, a is the _____ and bi is the imaginary part.

Example	Student Practice
1. Simplify.	**2.** Simplify.
(a) $\sqrt{-36}$	**(a)** $\sqrt{-41}$
For all positive real numbers a, $\sqrt{-a} = \sqrt{-1}\sqrt{a} = i\sqrt{a}$.	
$\sqrt{-36} = \sqrt{-1}\sqrt{36} = i(6) = 6i$	
(b) $\sqrt{-17}$	**(b)** $\sqrt{-25}$
$\sqrt{-17} = \sqrt{-1}\sqrt{17} = i\sqrt{17}$	
3. Multiply. $\sqrt{-16}\cdot\sqrt{-25}$	**4.** Multiply. $\sqrt{-36}\cdot\sqrt{-9}$
Rewrite using the definition $\sqrt{-1} = i$.	
$\left(\sqrt{-16}\right)\left(\sqrt{-25}\right) = \left(i\sqrt{16}\right)\left(i\sqrt{25}\right)$ $= i^2(4)(5)$ $= -1(20) = -20$	

Vocabulary Answers: 1. conjugates 2. imaginary number 3. complex number 4. real part

Example	Student Practice
5. Find the real numbers x and y if $x + 3i\sqrt{7} = -2 + yi$. Two complex numbers $a + bi$ and $c + di$ are equal if and only if $a = c$ and $b = d$. The real parts must be equal, so x must be -2. The imaginary parts must also be equal, so y must be $3\sqrt{7}$.	**6.** Find the real numbers x and y if $-3 + 3yi = x + 12i\sqrt{2}$.
7. Subtract. $(6 - 2i) - (3 - 5i)$ Subtract the real parts and subtract the imaginary parts. $(6 - 2i) - (3 - 5i)$ $= (6 - 3) + \left[-2 - (-5)\right]i$ $= 3 + (-2 + 5)i$ $= 3 + 3i$	**8.** Subtract. $(-5 + 6i) - (-2 + 4i)$
9. Multiply. $(7 - 6i)(2 + 3i)$ Use FOIL. $(7 - 6i)(2 + 3i)$ $= (7)(2) + (7)(3i) - (6i)(2) - (6i)(3i)$ $= 14 + 21i - 12i - 18i^2$ Since $i^2 = -1$, $14 + 21i - 12i - 18i^2$ can be written as $14 + 21i - 12i - 18(-1)$. Simplify this expression. $14 + 21i - 12i - 18(-1)$ $= 14 + 21i - 12i + 18$ $= 32 + 9i$ Thus, $(7 - 6i)(2 + 3i) = 32 + 9i$.	**10.** Multiply. $(6 - 5i)(3 - 2i)$

248

Example	Student Practice
11. Multiply. $3i(4-5i)$	**12.** Multiply. $-4i(3-6i)$

Use the distributive property. Recall that $i^2 = 1$.

$$3i(4-5i) = (3)(4)i + (3)(-5)i^2$$
$$= 12i - 15i^2$$
$$= 12i - 15(-1)$$
$$= 15 + 12i$$

13. Evaluate.	**14.** Evaluate.
(a) i^{36}	**(a)** i^{22}

Some values for i^n are shown below.

$$i = i \qquad i^5 = i \qquad i^9 = i$$
$$i^2 = -1 \quad i^6 = -1 \quad i^{10} = -1$$
$$i^3 = -i \quad i^7 = -i \quad i^{11} = -i$$
$$i^4 = +1 \quad i^8 = +1 \quad i^{12} = +1$$

Divide the exponent by 4 since i^4 raised to any power will be 1. Then use the values from the table above to evaluate the remainder.

$$i^{36} = \left(i^4\right)^9 = (1)^9 = 1$$

(b) i^{64}

(b) i^{27}

$$i^{27} = \left(i^{24+3}\right) = \left(i^{24}\right)\left(i^3\right)$$
$$= \left(i^4\right)^6 \left(i^3\right)$$
$$= (1)^6 (-i)$$
$$= -i$$

Example	Student Practice
15. Divide. $\dfrac{3-2i}{4i}$	**16.** Divide. $\dfrac{6-5i}{4+3i}$

Multiply the numerator and denominator by the conjugate of the denominator. The conjugate of $0+4i$ is $0-4i$, or simply $-4i$.

$$\frac{3-2i}{4i} = \frac{3-2i}{4i} \cdot \frac{-4i}{-4i}$$

$$= \frac{-12i+8i^2}{-16i^2}$$

$$= \frac{-12i+8(-1)}{-16(-1)}$$

$$= \frac{-8-12i}{16}$$

$$= \frac{\cancel{4}(-2-3i)}{\cancel{4}\cdot 4}$$

$$= \frac{-2-3i}{4} \text{ or } -\frac{1}{2}-\frac{3}{4}i$$

Extra Practice

1. Multiply and simplify. Place in i notation before doing any other operations.
$$\left(6+\sqrt{-2}\right)\left(3-\sqrt{-2}\right)$$

2. Evaluate. $i^{82}-i^{31}$

3. Multiply and simplify.
$$(3+2i)(2+i)$$

4. Divide. $\dfrac{6}{5-2i}$

Concept Check

Explain how you would simplify the following. $(3+5i)^2$

Name: _____   Date: _____

Instructor: _____   Section: _____

Chapter 8 Rational Exponents
8.7 Variation

Vocabulary

direct variation   •   inverse variation   •   combined variation   •   constant of variation

1. _____ is where one variable is a constant multiple of the reciprocal of the other.

2. In the direct variation equation $y = kx$, k is the _____.

3. _____ is where one variable is a constant multiple of the other.

4. If a quantity depends on the variation of two or more variables, it is called joint or _____.

Example	Student Practice
1. The time of a pendulum's period varies directly with the square root of its length. If the pendulum is 1 foot long when the time is 0.2 second, find the time when its length is 4 feet.	**2.** A car's stopping distance varies directly with the square of its speed. A car that is traveling 35 miles per hour can stop in 49 feet. What distance will it take to stop if it is traveling 60 miles per hour?

Let $t =$ the time and $L =$ the length.

This gives us the equation $t = k\sqrt{L}$. Substitute $L = 1$ and $t = 0.2$ into the equation to find k.

$$t = k\sqrt{L}$$
$$(0.2) = k\left(\sqrt{1}\right)$$
$$0.2 = k$$

We can rewrite the equation as $t = 0.2\sqrt{L}$. Find the value of t when $L = 4$.

$$t = 0.2\sqrt{L}$$
$$= 0.2\sqrt{4} = (0.2)(2) = 0.4 \text{ seconds}$$

Vocabulary Answers: 1. inverse variation 2. constant of variation 3. direct variation 4. combined variation

Example	Student Practice
3. If y varies inversely with x and $y = 12$ when $x = 5$, find the value of y when $x = 14$. Write the equation as $y = \dfrac{k}{x}$. Substitute $y = 12$ and $x = 5$ to find k. $12 = \dfrac{k}{5}$ $60 = k$ Rewrite the equation as $y = \dfrac{60}{x}$. Find the value of y when $x = 14$. $y = \dfrac{60}{x} = \dfrac{60}{14} = \dfrac{30}{7}$	**4.** If y varies inversely with x and $y = 12$ when $x = 5$, find the value of y when $x = 14$.
5. The amount of light from a light source varies inversely with the square of the distance from the light source. If an object receives 6.25 lumens when the light source is 8 meters away, how much light will the object receive if the light source is 4 meters away? Let $L =$ light and $d =$ distance, which gives the equation $L = \dfrac{k}{d^2}$. Substitute $L = 6.25$ and $d = 8$ to find k. $6.25 = \dfrac{k}{8^2}$ $400 = k$ Rewrite the original equation with $k = 400$. Then find L when $d = 4$. $L = \dfrac{400}{d^2} = \dfrac{400}{4^2} = \dfrac{400}{16} = 25$ lumens The check is left to the student.	**6.** The amount of light from a light source varies inversely with the square of the distance from the light source. If an object receives 9.31 lumens when the light source is 10 meters away, how much light will the object receive if the light source is 7 meters away?

Example	Student Practice
7. y varies directly with x and z and inversely with d^2. When $x = 7$, $z = 3$, and $d = 4$, the value of y is 20. Find the value of y when $x = 5$, $z = 6$, and $d = 2$.	**8.** y varies directly with m^4 and n and inversely with p^2. When $m = 2$, $n = 3$, and $p = 5$, the value of y is 63. Find the value of y when $m = 1$, $n = 8$, and $p = 3$.

Write the equation as $y = \dfrac{kxz}{d^2}$.

Substitute the known values and solve for k.

$$y = \frac{kxz}{d^2}$$

$$20 = \frac{k(7)(3)}{4^2}$$

$$20 = \frac{21k}{16}$$

$$320 = 21k$$

$$\frac{320}{21} = k$$

Substitute this value of k into the original equation.

$$y = \frac{\frac{320}{21}xz}{d^2} \quad \text{or} \quad y = \frac{320xz}{21d^2}$$

Now find y when $x = 5$, $z = 3$, and $d = 2$.

$$y = \frac{320xz}{21d^2}$$

$$y = \frac{320(5)(6)}{21(2)^2} = \frac{9600}{84} = \frac{800}{7}$$

Thus, the value of y is $\dfrac{800}{7}$.

Extra Practice

1. The distance a spring stretches varies directly with the weight of the object hung on the spring. If a 12-pound weight stretches a spring 15 inches, how far will a 32-pound weight stretch this spring?

2. If y varies inversely with the cube of x, and $y = 50$ when $x = 3$, find y when $x = 7$. Round to the nearest tenth.

3. The speed of a car varies inversely with the amount of time it takes to cover a certain distance. At 50 mph, a car travels a certain distance in 16 seconds. What is the speed of the car that travels the same distance in 10 seconds?

4. y varies directly with x, and inversely with z. $y = 40$ when $x = 6$ and $z = 15$. Find the value of y when $x = 0.5$ and $z = 12$. Round to the nearest tenth.

Concept Check

If y varies directly with the square root of x and $y = 50$ when $x = 5$, explain how you would find the constant of variation k.

MATH COACH

Mastering the skills you need to do well on the test.

Watch the **MATH COACH** videos in MyMathLab® or on YouTube™ while you work the problems below. These helpful hints will help you avoid making common errors on test problems.

Simplifying Expressions with Rational Exponents—
Problem 7   Evaluate. $16^{5/4}$

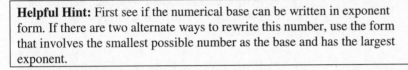

Helpful Hint: First see if the numerical base can be written in exponent form. If there are two alternate ways to rewrite this number, use the form that involves the smallest possible number as the base and has the largest exponent.

Did you write 16 as 2^4 ?

Yes _____   No _____

If you answered No, think of the different ways that 16 can be written in exponent form: 16^1, 4^2, or 2^4.

Your choice is most likely between 4^2 and 2^4. The problem works out more easily with the smallest possible base, which is 2. See if you can redo this step correctly.

Do you see that if we use the rule $\left(x^m\right)^n = x^{mn}$, we obtain

the expression $\left(2^4\right)^{5/4} = 2^{4/1 \cdot 5/4}$?

Yes _____   No _____

If you answered No, please review the rule about raising a power to a power and see if you can obtain the right result. Now simplify this expression further and evaluate the resulting exponential expression.

If you answered Problem 7 incorrectly, go back and rework the problem using these suggestions.

Adding and Subtracting Radical Expressions—Problem 12

Combine where possible. $\sqrt{40x} - \sqrt{27x} + 2\sqrt{12x}$

Helpful Hint: Remember to simplify each radical expression first. For this problem, remember to keep the variable x inside the radical.

Did you simplify $\sqrt{40x}$ to $2\sqrt{10x}$? Yes _____   No _____
Did you simplify $\sqrt{27x}$ to $3\sqrt{3x}$? Yes _____   No _____

If you answered No to either question, remember that $\sqrt{40x} = \sqrt{4} \cdot \sqrt{10x}$ and $\sqrt{27x} = \sqrt{9} \cdot \sqrt{3x}$. You can take the square roots of both 4 and 9. Try to do that part of the problem again and then simplify to see if you obtain the same results.

Did you simplify $2\sqrt{12x}$ to $4\sqrt{3x}$? Yes _____   No _____

If you answered No, remember that $2\sqrt{12x} = 2 \cdot \sqrt{4} \cdot \sqrt{3x}$. Go back and see if you can obtain the correct answer to that step. Remember that in your final step, you can only add radical expressions if they have the same radicand.

Now go back and rework the problem using these suggestions.

Rationalizing the Denominator—Problem 18   Rationalize the denominator. $\dfrac{1+2\sqrt{3}}{3-\sqrt{3}}$

Helpful Hint: You will need to multiply both the numerator and the denominator by the conjugate of the denominator. Do this very carefully using the FOIL method.

Did you multiply the numerator and the denominator by $\left(3+\sqrt{3}\right)$?

Yes _____   No _____

If you answered No, review the definition of conjugate. Note that the conjugate of $3-\sqrt{3}$ is $3+\sqrt{3}$.

Did you multiply the binomials in the denominator to get $9+3\sqrt{3}-3\sqrt{3}-\sqrt{9}$?

Yes _____   No _____

Did you multiply the binomials in the numerator to get $3+\sqrt{3}+6\sqrt{3}+2\sqrt{9}$?

Yes _____   No _____

If you answered No to either question, go back over each step of multiplying the binomial expressions using the FOIL method. Be careful in writing which numbers go outside the radical and which numbers go inside the radical. Now simplify each expression.

If you answered Problem 18 incorrectly, go back and rework the problem using these suggestions.

Solving a Radical Equation—Problem 20   Solve and check your solution(s). $5+\sqrt{x+15}=x$

Helpful Hint: Always isolate a radical term on one side of the equation before squaring each side.

Did you isolate the radical term first to obtain the equation $\sqrt{x+15}=x-5$?
Yes _____   No _____

After squaring each side, did you obtain $x+15=x^2-10x+25$?
Yes _____   No _____

If you answered No to either question, review the process for solving a radical equation. Remember that $\left(\sqrt{x+15}\right)^2=x+15$ and $(x-5)^2=x^2-10x+25$. Go back and try to complete these steps again.

Did you collect all like terms on one side to obtain the equation $0=x^2-11x+10$?
Yes _____   No _____

If you answered No, remember to add $-x-15$ to both sides of the equation.

Now factor the quadratic equation and set each factor equal to 0 to find the solution(s). Make sure to check your possible solutions and discard any solution that does not check when substituted back into the original equation.

If you answered Problem 20 incorrectly, go back and rework the problem using these suggestions.

Chapter 9 Quadratic Equations and Inequalities
9.1 Quadratic Equations

Vocabulary

quadratic equation • standard form • square root property • complete the square

1. The _____ states that if $x^2 = a$, then $x = \pm\sqrt{a}$ for all real numbers a.

2. When we _____ we are changing the polynomial to a perfect square trinomial.

3. $ax^2 + bx + c = 0$ is the _____ of a quadratic equation.

4. An equation written in the form $ax^2 + bx + c = 0$, where a, b, and c are real numbers and $a \neq 0$, is called a _____.

Example	Student Practice
1. Solve and check. $x^2 - 36 = 0$	**2.** Solve and check. $x^2 - 25 = 0$

1. Solve and check. $x^2 - 36 = 0$

If we add 36 to each side, we have $x^2 = 36$.

$x = \pm\sqrt{36}$
$x = \pm 6$

The two roots are 6 and −6.

Check.

$$(6)^2 - 36 \overset{?}{=} 0 \qquad (-6)^2 - 36 \overset{?}{=} 0$$
$$36 - 36 \overset{?}{=} 0 \qquad \quad 36 - 36 \overset{?}{=} 0$$
$$0 = 0 \qquad \qquad \quad 0 = 0$$

3. Solve. $x^2 = 48$

$x = \pm\sqrt{48}$
$x = \pm\sqrt{16 \cdot 3} = \pm 4\sqrt{3}$

The two roots are $4\sqrt{3}$ and $-4\sqrt{3}$.

4. Solve. $x^2 = 24$

Vocabulary Answers: 1. square root property 2. complete the square 3. standard form 4. quadratic equation

Example	Student Practice
5. Solve and check. $3x^2 + 2 = 77$	**6.** Solve and check. $6x^2 - 3 = 93$

5. $3x^2 + 2 = 77$

$$3x^2 + 2 = 77$$
$$3x^2 = 75$$
$$x^2 = 25$$
$$x = \pm\sqrt{25}$$
$$x = \pm 5$$

The roots are 5 and −5. The check is left to the student.

---

7. Solve and check. $4x^2 = -16$

$$4x^2 = -16$$
$$x^2 = -4$$
$$x = \pm\sqrt{-4}$$
$$x = \pm 2i$$

The roots are $2i$ and $-2i$.

Check.

$$4(2i)^2 \overset{?}{=} -16 \qquad 4(-2i)^2 \overset{?}{=} -16$$
$$4(-4) \overset{?}{=} -16 \qquad 4(-4) \overset{?}{=} -16$$
$$-16 = -16 \qquad -16 = -16$$

8. Solve and check. $2x^2 = -50$

---

9. Solve. $(4x-1)^2 = 5$

$$(4x-1)^2 = 5$$
$$4x - 1 = \pm\sqrt{5}$$
$$4x = 1 \pm \sqrt{5}$$
$$x = \frac{1 \pm \sqrt{5}}{4}$$

The roots are $\dfrac{1+\sqrt{5}}{4}$ and $\dfrac{1-\sqrt{5}}{4}$.

10. Solve. $(3x+8)^2 = 6$

Example	Student Practice
11. Solve by completing the square and check. $x^2 + 6x + 1 = 0$	**12.** Solve by completing the square and check. $x^2 + 4x + 1 = 0$

First rewrite the equation in the form $ax^2 + bx = c$. Thus, we obtain $x^2 + 6x = -1$.

Next verify that the coefficient of the quadratic term $\left(x^2\right)$ is 1. We want to add a constant term to $x^2 + 6x$ so that we get a perfect square trinomial. We do this by taking half the coefficient of x and squaring it.

$$\left(\frac{6}{2}\right)^2 = 3^2 = 9$$

Adding 9 to $x^2 + 6x$ gives the perfect square trinomial $x^2 + 6x + 9$, which we factor as $\left(x + 3\right)^2$. Add 9 to both the left and right side of our equation. We now have $x^2 + 6x + 9 = -1 + 9$.

Factor the left side then use the square root property.

$$\left(x + 3\right)^2 = 8$$
$$x + 3 = \pm\sqrt{8}$$
$$x + 3 = \pm 2\sqrt{2}$$

Next we solve for x by subtracting 3 from each side of the equation.

$$x = -3 \pm 2\sqrt{2}$$

The roots are $-3 + 2\sqrt{2}$ and $-3 - 2\sqrt{2}$.

The check is left to the student. Be sure to check the solution in the original equation, and not the perfect square trinomial.

Example	Student Practice
13. Solve by completing the square. $3x^2 - 8x + 1 = 0$	**14.** Solve by completing the square. $5x^2 - 4x - 3 = 0$

$3x^2 - 8x + 1 = 0$

$3x^2 - 8x = -1$

Divide each term by 3 so that the coefficient of the quadratic term is 1.

$$\frac{3x^2}{3} - \frac{8x}{3} = -\frac{1}{3}$$

$$x^2 - \frac{8}{3}x + \frac{16}{9} = -\frac{1}{3} + \frac{16}{9}$$

$$\left(x - \frac{4}{3}\right)^2 = \frac{13}{9}$$

$$x - \frac{4}{3} = \pm\sqrt{\frac{13}{9}}$$

$$x - \frac{4}{3} = \pm\frac{\sqrt{13}}{3}$$

$$x = \frac{4 \pm \sqrt{13}}{3}$$

Extra Practice

1. Solve the equation by using the square root property. Express any complex numbers using i notation. $x^2 + 64 = 0$

2. Solve the equation by using the square root property. Express any complex numbers using i notation. $(3x + 1)^2 = 15$

3. Solve the equation by completing the square. Express any complex numbers using i notation. $x^2 + 12x + 35 = 0$

4. Solve the equation by completing the square. Express any complex numbers using i notation. $4x^2 + 3 = x$

Concept Check

Explain how you would decide what to add to each side of the equation to complete the square for the following equation. $x^2 + x = 1$

Chapter 9 Quadratic Equations and Inequalities
9.2 The Quadratic Formula and Solutions to Quadratic Equations

Vocabulary

quadratic formula   •   standard form   •   discriminant   •   complex number

1. The expression $b^2 - 4ac$ is called the _____.

2. The _____ says that for all equations $ax^2 + bx + c = 0$, $x = \dfrac{-b \pm \sqrt{b^2 - 4ac}}{2a}$.

Example	Student Practice
1. Solve by using the quadratic formula. $$x^2 + 8x = -3$$ The standard form is $x^2 + 8x + 3 = 0$. We substitute $a = 1$, $b = 8$, and $c = 3$. $$x = \frac{-b \pm \sqrt{b^2 - 4ac}}{2a}$$ $$x = \frac{-8 \pm \sqrt{8^2 - 4(1)(3)}}{2(1)}$$ $$x = \frac{-8 \pm 2\sqrt{13}}{2}$$ $$x = -4 \pm \sqrt{13}$$	**2.** Solve by using the quadratic formula. $$x^2 + 4x = -7 - 8x$$
3. Solve by using the quadratic formula. $$2x^2 - 48 = 0$$ The standard form is $2x^2 - 0x - 48 = 0$. Therefore, $a = 2$, $b = 0$, and $c = -48$. $$x = \frac{-0 \pm \sqrt{(0)^2 - 4(2)(-48)}}{2(2)}$$ $$x = \frac{\pm 8\sqrt{6}}{4}$$ $$x = \pm 2\sqrt{6}$$	**4.** Solve by using the quadratic formula. $$4x^2 - 60 = 0$$

Vocabulary Answers: 1. discriminant 2. quadratic formula

Example	Student Practice

5. A small company that manufactures canoes makes a daily profit p according to the equation $p = -100x^2 + 3400x - 26,196$, where p is measured in dollars and x is the number of canoes made per day. Find the number of canoes that must be made each day to produce a zero profit for the company. Round your answer to the nearest whole number.

Since $p = 0$, the equation is $0 = -100x^2 + 3400x - 26,196$. Thus, $a = -100$, $b = 3400$, and $c = -26,196$.

$$x = \frac{-3400 \pm \sqrt{3400^2 - 4(-100)(-26,196)}}{2(-100)}$$

$$x = \frac{-3400 \pm \sqrt{1,081,600}}{-200}$$

$$x = \frac{-3400 \pm 1040}{-200}$$

We now obtain two answers.

$$x = \frac{-3400 + 1040}{-200}$$

$$x = \frac{-2360}{-200} = 11.8 \approx 12$$

$$x = \frac{-3400 - 1040}{-200}$$

$$x = \frac{-4440}{-200} = 22.2 \approx 22$$

A zero profit is obtained when approximately 12 canoes are produced or when approximately 22 canoes are produced.

6. A company that manufactures guitars makes a daily profit p according to the equation $p = -100x^2 + 3024x - 20,240$, where p is measured in dollars and x is the number of guitars made per day. Find the number of guitars that must be made each day to produce a zero profit for the company. Round your answer to the nearest whole number.

Example	Student Practice
7. Solve by using the quadratic formula. $$\frac{2x}{x+2} = 1 - \frac{3}{x+4}$$ The LCD is $(x+2)(x+4)$. $$\frac{2x}{x+2} = 1 - \frac{3}{x+4}$$ $$2x(x+4) = (x+2)(x+4) - 3(x+2)$$ $$x^2 + 5x - 2 = 0$$ $$x = \frac{-5 \pm \sqrt{5^2 - 4(1)(-2)}}{2(1)}$$ $$x = \frac{-5 \pm \sqrt{33}}{2}$$	**8.** Solve by using the quadratic formula. $$\frac{6}{x} + \frac{x}{x-3} = -\frac{4}{5}$$
9. Solve and simplify your answer. $$8x^2 - 4x + 1 = 0$$ $$x = \frac{-(-4) \pm \sqrt{(-4)^2 - 4(8)(1)}}{2(8)}$$ $$x = \frac{4 \pm 4i}{16}$$ $$x = \frac{1 \pm i}{4}$$	**10.** Solve by using the quadratic formula. $$6x^2 + 4x + 3 = 0$$
11. What type of solutions does the equation $2x^2 - 9x - 35 = 0$ have? Do not solve the equation. $a = 2$, $b = -9$, and $c = -35$. $$b^2 - 4ac = \left(-9^2\right) - 4(2)(-35) = 361$$ Since the discriminant is positive, the equation has two real roots. 361 is a perfect square. The equation has two different rational solutions.	**12.** Use the discriminant to find what type of solutions the equation $2x^2 - 5x + 43 = 0$ has. Do not solve the equation.

Example	Student Practice
13. Find a quadratic equation whose roots are 5 and -2. $x = 5 \qquad x = -2$ $x - 5 = 0 \quad x + 2 = 0$ $(x - 5)(x + 2) = 0$ $x^2 - 3x - 10 = 0$	**14.** Find a quadratic equation whose roots are 3 and 4.
15. Find a quadratic equation those solutions are $3i$ and $-3i$. $x - 3i = 0 \quad \text{and} \quad x + 3i = 0$ $(x - 3i)(x + 3i) = 0$ $x^2 + 3ix - 3ix - 9i^2 = 0$ $x^2 + 9 = 0$	**16.** Find a quadratic equation whose solutions are $3i\sqrt{5}$ and $-3i\sqrt{5}$

Extra Practice

1. Solve by the quadratic formula. Simplify your answers. $3x^2 + 12x + 7 = 0$

2. Simplify, then solve by the quadratic formula. Simplify your answers. Use i notation for nonreal complex numbers.
$$\frac{1}{x} - \frac{2}{x + 2} = \frac{1}{5}$$

3. Use the discriminant to find what type of solutions the following equation has. Do not solve the equation.
$x^2 - 3(2x - 3) = 0$

4. Write a quadratic equation having the given solutions -3 and $-\dfrac{1}{2}$.

Concept Check

Explain how you would determine if $2x^2 - 6x + 3 = 3$ has two rational, two irrational, one rational, or two nonreal complex solutions.

Chapter 9 Quadratic Equations and Inequalities
9.3 Equations That Can Be Transformed into Quadratic Form

Vocabulary

quadratic in form   •   linear term   •   standard form   •   substitution

1. Before making a substitution and then factoring, we first must put the equation into _____.

2. An equation is _____ if we can substitute a linear term for the variable raised to the lowest power and get an equation of the form $ay^2 + by + c = 0$.

Example	Student Practice
1. Solve. $x^4 - 13x^2 + 36 = 0$	**2.** Solve. $x^4 + 9x^2 - 400 = 0$

1. Solve. $x^4 - 13x^2 + 36 = 0$

Let $y = x^2$. Then $y^2 = x^4$. Substitute these into the equation.

$y^2 - 13y + 36 = 0$

Factor the equation and then solve for x.

$(y-4)(y-9) = 0$

$y - 4 = 0$ or $y - 9 = 0$

    $y = 4$ $y = 9$

Replace y with x^2.

$x^2 = 4$ or $x^2 = 9$

$x = \pm\sqrt{4}$ $x = \pm\sqrt{9}$

$x = \pm 2$ $x = \pm 3$

Thus, there are four solutions to the original equation: $x = +2$, $x = -2$, $x = +3$, and $x = -3$. Check these values to verify that they are solutions.

The check is left to the student.

Vocabulary Answers: 1. standard form 2. quadratic in form

Example	Student Practice
3. Solve for all real roots. $2x^6 - x^3 - 6 = 0$	**4.** Solve for all real roots. $3x^6 - 83x^3 + 54 = 0$

3. (continued)

Let $y = x^3$. Then $y^2 = x^6$. Thus, we have the following:

$$2y^2 - y - 6 = 0$$
$$(2y+3)(y-2) = 0$$
$$2y+3 = 0 \qquad \text{or} \quad y-2 = 0$$
$$y = -\frac{3}{2} \qquad\qquad y = 2$$
$$x^3 = -\frac{3}{2} \qquad\qquad x^3 = 2$$
$$x = \sqrt[3]{-\frac{3}{2}} \qquad\qquad x = \sqrt[3]{2}$$
$$x = \frac{\sqrt[3]{-12}}{2}$$

The check is left to the student.

---

5. Solve and check your solutions.

$x^{2/3} - 3x^{1/3} + 2 = 0$

Let $y = x^{1/3}$. Then $y^2 = x^{2/3}$.

$$y^2 - 3y + 2 = 0$$
$$(y-2)(y-1) = 0$$
$$y-2 = 0 \quad \text{or} \quad y-1 = 0$$
$$y = 2 \qquad\qquad y = 1$$
$$x^{1/3} = 2 \qquad\qquad x^{1/3} = 1$$
$$\left(x^{1/3}\right)^3 = (2)^3 \qquad \left(x^{1/3}\right)^3 = (1)^3$$
$$x = 8 \qquad\qquad x = 1$$

The check is left to the student.

6. Solve and check your solutions.

$x^{2/3} - 65x^{1/3} + 64 = 0$

Example	Student Practice
7. Solve and check your solutions. $2x^{1/2} = 5x^{1/4} + 12$	**8.** Solve and check your solutions. $2x^{1/2} = 11x^{1/4} + 15$

This equation in standard form is
$2x^{1/2} - 5x^{1/4} - 12 = 0$. Let $y = x^{1/4}$,
making $y^2 = x^{1/2}$. Then solve.

$$2y^2 - 5y - 12 = 0$$
$$(2y + 3)(y - 4) = 0$$
$$2y + 3 = 0 \qquad \text{or} \quad y - 4 = 0$$
$$y = -\frac{3}{2} \qquad\qquad y = 4$$
$$x^{1/4} = -\frac{3}{2} \qquad\quad x^{1/4} = 4$$
$$\left(x^{1/4}\right)^4 = \left(-\frac{3}{2}\right)^4 \quad \left(x^{1/4}\right)^4 = (4)^4$$
$$x = \frac{81}{16} \qquad\qquad x = 256$$

Check by substituting the solutions into
the original equation.

$$x = \frac{81}{16}$$
$$2\left(\frac{81}{16}\right)^{1/2} - 5\left(\frac{81}{16}\right)^{1/4} - 12 \overset{?}{=} 0$$
$$\frac{9}{2} - \frac{15}{2} - 12 \overset{?}{=} 0$$
$$-15 \neq 0$$
$$x = 256$$
$$2(256)^{1/2} - 5(256)^{1/4} - 12 \overset{?}{=} 0$$
$$32 - 20 - 12 \overset{?}{=} 0$$
$$0 = 0$$

$\dfrac{81}{16}$ is extraneous and not a valid
solution. The only valid solution is 256.

Example	Student Practice
9. Solve and check your solutions. $x^{-2} = 5x^{-1} + 14$	**10.** Solve and check your solutions. $x^{-2} + x^{-1} = 12$

Write the equation in standard form. Let $y = x^{-1}$, making $y^2 = x^{-2}$. Then solve.

$$x^{-2} = 5x^{-1} + 14$$
$$x^{-2} - 5x^{-1} - 14 = 0$$
$$y^2 - 5y - 14 = 0$$
$$(y - 7)(y + 2) = 0$$
$$y - 7 = 0 \quad \text{or} \quad y + 2 = 0$$
$$y = 7 \qquad \qquad y = -2$$
$$x^{-1} = 7 \qquad \quad x^{-1} = -2$$
$$x = \frac{1}{7} \qquad \qquad x = -\frac{1}{2}$$

The check is left to the student.

Extra Practice

1. Solve. Express any nonreal complex numbers with i notation.
$$x^4 + 10x^2 + 21 = 0$$

2. Solve for all real roots. $x^6 + 27x^3 = 0$

3. Solve for all real roots.
$$6x^{2/5} + 18x^{1/5} + 12 = 0$$

4. Solve for all real roots. $3x^{-2} + 3x^{-1} = 168$

Concept Check

Explain how you would solve the following. $x^8 - 6x^4 = 0$

Chapter 9 Quadratic Equations and Inequalities
9.4 Formulas and Applications

Vocabulary

hypotenuse   •   leg   •   Pythagorean Theorem   •   area   •   surface area

1. The _____ states that if c is the length of the longest side of a right triangle and a and b are the lengths of the other two sides, then $a^2 + b^2 = c^2$.

2. The longest side of a right triangle is called the _____.

Example	**Student Practice**
1. The surface area of a sphere is given by $A = 4\pi r^2$. Solve this equation for r. (You do not need to rationalize the denominator.) $$A = 4\pi r^2$$ $$\frac{A}{4\pi} = r^2$$ $$\pm\sqrt{\frac{A}{4\pi}} = r$$ $$\pm\frac{1}{2}\sqrt{\frac{A}{\pi}} = r$$ Since the radius of a sphere must be a positive value, we use only the principal root. $r = \frac{1}{2}\sqrt{\frac{A}{\pi}}$	**2.** The volume of a sphere is given by $V = \frac{4}{3}\pi r^3$. Solve this equation for r. (You do not need to rationalize the denominator.)
3. Solve for y. $y^2 - 2yz - 15z^2 = 0$ Factor, then set each factor equal to 0. $$y^2 - 2yz - 15z^2 = 0$$ $$(y + 3z)(y - 5z) = 0$$ $$y + 3z = 0 \quad \text{or} \quad y - 5z = 0$$ $$y = -3z \qquad\quad y = 5z$$	**4.** Solve for x. $x^2 - 3xy - 10y^2 = 0$

Vocabulary Answers: 1. Pythagorean Theorem 2. hypotenuse

Example	Student Practice

5. Solve for x. $2x^2 + 3wx - 4z = 0$

We use the quadratic formula where the variable is considered to be x, and w and z are considered constants. Thus, $a = 2$, $b = 3w$, and $c = -4z$.

$$x = \frac{-b \pm \sqrt{b^2 - 4ac}}{2a}$$

$$x = \frac{-3w \pm \sqrt{(3w)^2 - 4(2)(-4z)}}{2(2)}$$

$$x = \frac{-3w \pm \sqrt{9w^2 + 32z}}{4}$$

6. Solve for x. $4x^2 - 6yx + 9z = 0$

7. Complete parts **(a)** and **(b)**.

(a) Solve the Pythagorean Theorem $a^2 + b^2 = c^2$ for a.

$$a^2 + b^2 = c^2$$
$$a^2 = c^2 - b^2$$
$$a = \pm\sqrt{c^2 - b^2}$$

a, b, and c must be positive numbers because they represent lengths, only use the positive root, $a = \sqrt{c^2 - b^2}$.

(b) Find the value of a if $c = 13$ and $b = 5$.

$$a = \sqrt{c^2 - b^2}$$
$$a = \sqrt{(13)^2 - (5)^2}$$
$$a = \sqrt{144}$$
$$a = 12$$

8. Complete parts **(a)** and **(b)**.

(a) Solve the Pythagorean Theorem $a^2 + b^2 = c^2$ for c.

(b) Find the value of c if $a = 15$ and $b = 20$.

Example	Student Practice
9. The perimeter of a triangular piece of land is 12 miles. One leg of the triangle is 1 mile longer than the other leg. Find the length of each boundary of the land if the triangle is a right triangle.	**10.** The perimeter of a triangular piece of land is 60 miles. One leg of the triangle is 4 miles longer than twice the other leg. Find the length of each boundary of the land if the triangle is a right triangle.

We are given that the perimeter is 12 miles, so $x + (x+1) + c = 12$.

Thus, $c = -2x + 11$.

By the Pythagorean Theorem,
$x^2 + (x+1)^2 = (-2x+11)^2$.

$$x^2 + (x+1)^2 = (-2x+11)^2$$
$$x^2 + x^2 + 2x + 1 = 4x^2 - 44x + 121$$
$$0 = 2x^2 - 46x + 120$$
$$0 = x^2 - 23x + 60$$

Use the quadratic formula to solve.
$$x = \frac{-(23) \pm \sqrt{(-23)^2 - 4(1)(60)}}{2(1)}$$

$$x = \frac{23 \pm \sqrt{289}}{2}$$

$$x = \frac{23 \pm 17}{2}$$

$$x = \frac{40}{2} = 20 \text{ or } x = \frac{6}{2} = 3.$$

20 is too large, so the only answer that makes sense is $x = 3$.

Thus, the sides of the triangle are $x = 3$, $x + 1 = 4$, and $-2x + 11 = -2(3) + 11 = 5$.

The longest boundary of this triangular piece of land is 5 miles. The other two boundaries are 4 miles and 3 miles.

Example	Student Practice
11. A triangular sign marks the edge of the rocks in Rockport Harbor. The sign has an area of 35 square meters. Find the base and altitude of this triangular sign if the base is 3 meters shorter than the altitude.	**12.** The length of a rectangle is 2 yards shorter than twice the width. The area of the rectangle is 84 square yards. Find the dimensions of the rectangle.

The area of the triangle is given by $A = \dfrac{1}{2}ab$. Let x = the length in meters of the altitude. Then $x - 3$ = the length in meters of the base. Solve for x.

$$35 = \frac{1}{2}x(x-3)$$
$$0 = x^2 - 3x - 70$$
$$0 = (x-10)(x+7)$$
$$x = 10 \text{ or } x = -7$$

We disregard -7. Thus altitude $= x = 10$ meters and base $= x - 3 = 7$ meters. The check is left to the student.

Extra Practice

1. Solve $s = \dfrac{1}{2}gt^2$ for t. Assume that all other variables are nonzero.

2. Solve $(a+2)x^2 - 7x + 3y = 0$ for x. Assume that all other variables are nonzero.

3. $c = 10$ and $b = 3a$; use the Pythagorean Theorem to find a and b.

4. Andrew drove at a constant speed on a dirt road for 75 miles. He then traveled 20 mph faster on a paved road for 180 miles. If he drove for 7 hours, find the car's speed for each part of the trip.

Concept Check

In a right triangle the hypotenuse measures 12 meters. One of the legs of the triangle is three times as long as the other. Explain how you would find the length of each leg.

272

Chapter 9 Quadratic Equations and Inequalities
9.5 Quadratic Functions

Vocabulary

vertex   •   quadratic function

1. A _____ is a function of the form $f(x) = ax^2 + bx + c$.

2. The _____ is the lowest point on a parabola opening upward or the highest point on a parabola opening downward.

Example	Student Practice
1. Find the coordinates of the vertex and the intercepts of the quadratic function $f(x) = x^2 - 8x + 15$.	2. Find the coordinates of the vertex and the intercepts of the quadratic function $g(x) = x^2 - 7x + 12$.

The vertex occurs at $\dfrac{-b}{2a} = \dfrac{-(-8)}{2(1)} = 4$.

To find the y-coordinate, find $f(4)$.

$$f(4) = 4^2 - 8(4) + 15$$
$$= 16 - 32 + 15 = -1$$

The vertex is $(4, -1)$. The y-intercept is at $f(0)$.

$$f(0) = 0^2 - 8(0) + 15 = 15$$

The y-intercept is $(0, 15)$. The x-intercepts occur when $x^2 - 8x + 15 = 0$. Solve for x.

$$(x - 5)(x - 3) = 0$$
$$x - 5 = 0 \quad x - 3 = 0$$
$$x = 5 \qquad x = 3$$

The x-intercepts are $(5, 0)$ and $(3, 0)$.

Vocabulary Answers: 1. quadratic function 2. vertex

Example	Student Practice
3. Find the vertex and the intercepts, and then graph the function $f(x) = x^2 + 2x - 4$.	**4.** Find the vertex and the intercepts, and then graph the function $f(x) = x^2 - 4x + 1$.

Since $a > 0$, the parabola opens upward. Find the vertex.

$$x = \frac{-b}{2a} = \frac{-2}{2(1)} = \frac{-2}{2} = -1$$

$$f(-1) = (-1)^2 + 2(-1) - 4 = -5$$

The vertex is $(-1, -5)$. The y-intercept is at $f(0)$.

$$f(0) = (0)^2 + 2(0) - 4 = -4$$

The y-intercept is $(0, -4)$. The x-intercepts occur when $f(x) = 0$. We use the quadratic formula.

$$x = \frac{-b \pm \sqrt{b^2 - 4ac}}{2a} = \frac{-2 \pm \sqrt{2^2 - 4(1)(-4)}}{2(1)}$$

$$= -1 \pm \sqrt{5}$$

The x-intercepts are approximately $(-3.2, 0)$ and $(1.2, 0)$. The vertex is $(-1, -5)$; the y-intercept is $(0, -4)$; and the x-intercepts are approximately $(-3.2, 0)$ and $(1.2, 0)$. Connect these points by a smooth curve to graph the parabola.

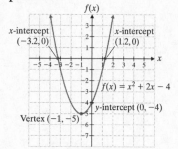

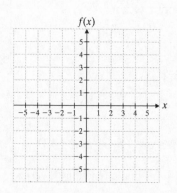

274

Example	Student Practice
5. Find the vertex and the intercepts, and then graph the function $f(x) = -2x^2 + 4x - 3$.	**6.** Find the vertex and the intercepts, and then graph the function $f(x) = -3x^2 - 6x - 6$.

Since $a < 0$, the parabola opens downward. Find the vertex.

$$x = \frac{-4}{2(-2)} = \frac{-4}{-4} = 1$$

$$f(1) = -2(1)^2 + 4(1) - 3 = -1$$

The vertex is $(1, -1)$. The y-intercept is at $f(0)$. $f(0) = -2(0)^2 + 4(0) - 3$ $= -3$. The y-intercept is $(0, -3)$. The x-intercepts occur when $f(x) = 0$. We use the quadratic formula.

$$x = \frac{-4 \pm \sqrt{4^2 - 4(-2)(-3)}}{2(-2)}$$

$$= \frac{-4 \pm \sqrt{-8}}{-4}$$

Because $\sqrt{-8}$ yields an imaginary number, there are no x-intercepts for the graph of the function. We will look for three additional points. We try $f(2)$, $f(3)$, and $f(-1)$ and get the points $(2, -3)$, $(3, -9)$, and $(-1, -9)$.

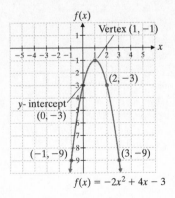

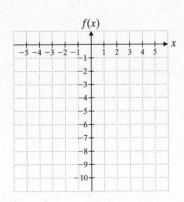

Extra Practice

1. Find the coordinates of the vertex and the intercepts of $g(x) = x^2 + 5x - 6$. When necessary, approximate x-intercepts to the nearest tenth.

2. Find the coordinates of the vertex and the intercepts of $f(x) = 5x^2 + 17x - 12$. When necessary, approximate x-intercepts to the nearest tenth.

3. Find the vertex, the y-intercept, and the x-intercepts (if any exist), and then graph the function $r(x) = 3x^2 - 2x + 1$.

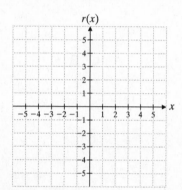

4. Find the vertex, the y-intercept, and the x-intercepts (if any exist), and then graph the function $f(x) = -x^2 + 2x - 1$.

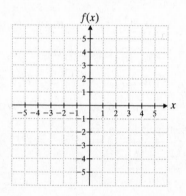

Concept Check

Explain how you would find the vertex of $f(x) = 4x^2 - 9x - 5$.

Chapter 9 Quadratic Equations and Inequalities
9.6 Compound and Quadratic Inequalities in One Variable

Vocabulary

quadratic inequality   •   boundary points   •   compound inequalities

empty set   •   and   •   or

1. The notation $\varnothing$ represents the _____.

2. A(n) _____ has the form $ax^2 + bx + c < 0$ (or replace $<$ by $>$, $\leq$, or $\geq$), where a, b, and c are real numbers and $a \neq 0$.

3. Inequalities that consist of two inequalities connected by the word and or the word or are called _____.

4. The two points where the expression of a quadratic inequality is equal to zero are called _____.

Example	Student Practice
1. Graph the values of x where $7 < x$ and $x < 12$. We read the inequality starting with the variable. Thus, we graph all values of x, where x is greater than 7 and where x is less than 12. All such values must be between 7 and 12. Numbers that are greater than 7 and less than 12 can be written as $7 < x < 12$. 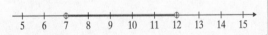	**2.** Graph the values of x where $-2 < x$ and $x < 2$.
3. Graph the region where $x < 3$ or $x > 6$. Read the inequality as "x is less than 3 or x is greater than 6." This includes all values to the left of 3 as well as all values to the right of 6 on a number line. We shade these regions. 	**4.** Graph the region where $x < -1$ or $x > 1$.

Vocabulary Answers: 1. empty set 2. quadratic inequality 3. compound inequality 4. boundary points

Example	Student Practice
5. Solve for x and graph the compound solution. $3x + 2 > 14$ or $2x - 1 < -7$	**6.** Solve for x and graph the compound solution. $4x - 6 \leq -2$ or $5x + 2 \geq 22$

We solve each inequality separately.

$$3x + 2 > 14 \quad \text{or} \quad 2x - 1 < -7$$
$$3x > 12 \qquad\qquad 2x < -6$$
$$x > 4 \qquad\qquad x < -3$$

The solution is $x < -3$ or $x > 4$.

7. Solve and graph $x^2 - 10x + 24 > 0$.

8. Solve and graph $x^2 - x - 12 > 0$.

Replace the inequality symbol with an equals sign and solve the resulting equation.

$$x^2 - 10x + 24 = 0$$
$$(x - 4)(x - 6) = 0$$
$$x - 4 = 0 \quad \text{or} \quad x - 6 = 0$$
$$x = 4 \qquad\qquad x = 6$$

We use the boundary points to separate the number line into distinct regions.

Evaluate the quadratic expression at a test point in each of the regions. Pick the test point $x = 1$ and get $15 > 0$. Pick the test point $x = 5$ and get $-1 < 0$. Pick the test point $x = 7$ and get $3 > 0$. Thus, $x^2 - 10x + 24 > 0$ when $x < 4$ or when $x > 6$.

278

Example	Student Practice

9. Solve and graph $x^2 + 4x > 6$. Round your answer to the nearest tenth.

First we write $x^2 + 4x - 6 > 0$. Because we cannot factor $x^2 + 4x - 6$, we use the quadratic formula to find the boundary points.

$$x = \frac{-4 \pm \sqrt{4^2 - 4(1)(-6)}}{2(1)}$$

$$= \frac{-4 \pm \sqrt{40}}{2}$$

$$= -2 \pm \sqrt{10}$$

$$-2 + \sqrt{10} \approx 1.2 \text{ or } -2 - \sqrt{10} \approx -5.2$$

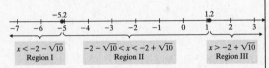

We will see where $x^2 + 4x - 6 > 0$. Test $x = -6$.

$$(-6)^2 + 4(-6) - 6 = 36 - 24 - 6 = 6 > 0$$

Test $x = 0$.

$$(0)^2 + 4(0) - 9 = 0 + 0 - 6 = -6 < 0$$

Test $x = 2$.

$$(2)^2 + 4(2) - 6 = 4 + 8 - 6 = 6 > 0$$

Our answer is $x < -5.2$ or $x > 1.2$.

10. Solve and graph $x^2 + 4x > 1$. Round your answer to the nearest tenth.

Extra Practice

1. Graph the values of x that satisfy the conditions given. $-4 < x \leq \dfrac{1}{2}$

2. Graph the values of x that satisfy the conditions given. $x \leq 4$ or $x \geq \dfrac{13}{2}$

3. Solve for x and graph your results.
$x - 3 \geq 2$ or $x + 1 \leq 2$

4. Solve and graph. $2x^2 - 3x - 5 \leq 0$

Concept Check

Explain what happens when you solve the inequality $x^2 + 2x + 8 > 0$.

Chapter 9 Quadratic Equations and Inequalities
9.7 Absolute Value Equations and Inequalities

Vocabulary

absolute value • $|a| = |b|$ • $|ax+b| = c$ $|ax+b| < c$ • $|ax+b| > c$

1. The _____ of a number x can be pictured as the distance between 0 and x on the number line.

2. _____ is equivalent to $ax+b < -c$ or $ax+b > c$.

Example	**Student Practice**
1. Solve and check your solutions. $\|2x+5\| = 11$	**2.** Solve and check your solutions. $\|4x+6\| = 18$

Example (continued)

The solutions of an equation of the form $|ax+b| = c$, where $a \neq 0$ and c is a positive number, are those values that satisfy $ax+b = c$ or $ax+b = -c$. Thus, we have the following:

$$2x+5 = 11 \quad \text{or} \quad 2x+5 = -11$$
$$2x = 6 \qquad\qquad 2x = -16$$
$$x = 3 \qquad\qquad x = -8$$

The two solutions are 3 and -8. Check the solutions.

$$|2x+5| = 11 \qquad\qquad |2x+5| = 11$$
$$|2(3)+5| \stackrel{?}{=} 11 \qquad |2(-8)+5| \stackrel{?}{=} 11$$
$$|6+5| \stackrel{?}{=} 11 \qquad\quad |-16+5| \stackrel{?}{=} 11$$
$$|11| \stackrel{?}{=} 11 \qquad\qquad |-11| \stackrel{?}{=} 11$$
$$11 = 11 \qquad\qquad\quad 11 = 11$$

The solutions check.

Vocabulary Answers: 1. absolute value 2. $|ax+b| > c$

Example	Student Practice
3. Solve $\|3x-1\|+2=5$ and check your solutions.	**4.** Solve $\|6x-3\|-4=5$ and check your solutions.

First rewrite the equation so that the absolute value expression is alone on one side of the equation.

$$\|3x-1\|+2=5$$

$$\|3x-1\|+2-2=5-2$$

$$\|3x-1\|=3$$

Now solve $\|3x-1\|=3$ for x.

$$3x-1=3 \quad \text{or} \quad 3x-1=-3$$

$$x=\frac{4}{3} \qquad\qquad x=-\frac{2}{3}$$

---

5. Solve and check. $\|3x-4\|=\|x+6\|$

6. Solve and check. $\|x+4\|=\|4x+6\|$

The solutions of the given equation must satisfy $3x-4=x+6$ or $3x-4=-(x+6)$. Solve each equation to get $x=5$ and $x=-\frac{1}{2}$.

Check the solutions by substituting them into the original equation.

$$x=5: \quad \|3(5)-4\|\overset{?}{=}\|5+6\|$$

$$\|11\|=\|11\|$$

$$x=-\frac{1}{2}: \quad \left\|3\left(-\frac{1}{2}\right)-4\right\|\overset{?}{=}\left\|-\frac{1}{2}+6\right\|$$

$$\left\|-\frac{3}{2}-4\right\|\overset{?}{=}\left\|-\frac{1}{2}+6\right\|$$

$$\left\|-\frac{3}{2}-\frac{8}{2}\right\|\overset{?}{=}\left\|-\frac{1}{2}+\frac{12}{2}\right\|$$

$$\frac{11}{2}=\frac{11}{2}$$

Example	Student Practice
7. Solve and graph the solution. $\left|x\right| \le 4.5$	**8.** Solve and graph. $\left|x\right| \le 3$

7. (continued)

The inequality $\left|x\right| \le 4.5$ means that x is less than or equal to 4.5 units from 0 on a number line. We draw a picture.

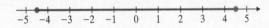

Thus, the solution is $-4.5 \le x \le 4.5$.

---

9. Solve and graph the solution. $\left|x+5\right| \le 10$	**10.** Solve and graph the solution. $\left|x+2\right| \le 3$

9. (continued)

We want to find the values of x that make $-10 \le x+5 \le 10$ a true statement. We need to solve the compound inequality. Subtract 5 from each part.

$$-10-5 \le x+5-5 \le 10-5$$
$$-15 \le x \le 5$$

Thus, the solution is $-15 \le x \le 5$. We graph this solution.

---

11. Solve and graph the solution. $\left|x\right| \ge 5\dfrac{1}{4}$	**12.** Solve and graph the solution. $\left|x\right| \ge 4$

11. (continued)

The inequality means that x is more than $5\dfrac{1}{4}$ units from 0 on a number line.

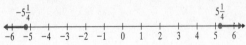

The solution is $x \le -5\dfrac{1}{4}$ or $x \ge 5\dfrac{1}{4}$.

Example	Student Practice
13. Solve and graph the solution. $$\left|-3x+6\right|>18$$	**14.** Solve and graph the solution. $$\left|-2x+5\right|\geq3$$

13. (continued)

Remember to reverse the inequality sign when dividing by a negative number.

$$-3x+6>18 \quad \text{or} \quad -3x+6<-18$$
$$-3x>12 \qquad\qquad -3x<-24$$
$$\frac{-3x}{-3}<\frac{12}{-3} \qquad \frac{-3x}{-3}>\frac{-24}{-3}$$
$$x<-4 \qquad\qquad x>8$$

Extra Practice

1. Solve. Check your solutions. $\left|2x+9\right|=31$

2. Solve. Check your solutions. $$\left|3m+2\right|-10=-6$$

3. Solve and graph the solution. $\left|x-3\right|\leq3$

4. Solve. In a certain company, the measured thickness t of a computer chip must not differ from the standard s by more than 0.05 millimeter. The engineers express this requirement as $\left|t-s\right|\leq0.05$. Find the limits of t if the standard s is 1.33 millimeters.

Concept Check

Explain what happens when you try to solve for x. $\left|7x+3\right|<-4$

MATH COACH

Mastering the skills you need to do well on the test.

Watch the MATH COACH videos in MyMathLab®or on You Tube while you work the problems below. These helpful hints will help you avoid making common errors on test problems.

Solving Quadratic Equations Involving Fractions—Problem 6

Solve the quadratic equation. 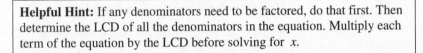 $\dfrac{2x}{2x+1} - \dfrac{6}{4x^2-1} = \dfrac{x+1}{2x-1}$

Helpful Hint: If any denominators need to be factored, do that first. Then determine the LCD of all the denominators in the equation. Multiply each term of the equation by the LCD before solving for x.

Did you factor $4x^2-1$ as $(2x+1)(2x-1)$?

Yes _____ No _____

Did you identify the LCD to be $(2x+1)(2x-1)$?

Yes _____ No _____

If you answered No to these questions, review how to factor the difference of two squares and how to find the LCD of polynomial denominators.

Did you multiply the LCD by each term of the equation and remove parentheses to obtain $2x^2-5x-7=0$?

Yes _____ No _____

Did you then use the quadratic formula and substitute for a,

b, and c to get $x = \dfrac{-(-5) \pm \sqrt{(-5)^2 - 4(2)(-7)}}{2(2)}$?

Yes _____ No _____

If you answered No to these questions, remember that your final equation should be in the form $ax^2+bx+c=0$. Then use $a=2$, $b=-5$, and $c=-7$ in the quadratic formula and simplify your result.

If you answered Problem 6 incorrectly, go back and rework the problem using these suggestions.

Solving Equations That Are Quadratic in Form—Problem 9

Solve for any valid real roots. $x^4 - 11x^2 + 18 = 0$

Helpful Hint: Let $y = x^2$ and then let $y^2 = x^4$. Write the new quadratic equation after these replacements have been made.

After making the necessary replacements, did you obtain the equation $y^2 - 11y + 18 = 0$? Yes _____ No _____

Did you solve the quadratic equation using any method to result in $y = 9$ and $y = 2$? Yes _____ No _____

If you answered No to these questions, stop and complete these steps again.

If $y = 9$ and $y = 2$, can you conclude that $x^2 = 9$ and $x^2 = 2$? Yes _____ No _____

If you take the square root of each side of each equation, can you obtain $x = \pm 3$ and $x = \pm\sqrt{2}$? Yes _____ No _____

If you answered No to these questions, remember that when you take the square root of each side of the equation there are two sign possibilities. Note that your final solution should consist of four values for x.

Now go back and rework the problem using these suggestions.

Solving a Quadratic Equation with Several Variables—Problem 13

Solve for the variable specified. $5y^2 + 2by + 6w = 0$; for y

> **Helpful Hint:** Think of the equation being written as $ay^2 + by + c = 0$. The quantities for a, b, or c may contain variables. Use the quadratic formula to solve for y.

Did you determine that $a = 5$, $b = 2b$, and $c = 6w$?

Yes ____ No ____

Did you substitute these values into the quadratic formula

and simplify to obtain $y = \dfrac{-2b \pm \sqrt{4b^2 - 120w}}{10}$?

Yes ____ No ____

If you answered No to these questions, review the Helpful Hint again to make sure that you find the correct values for a, b, and c. Carefully substitute these values into the quadratic formula and simplify.

Were you able to simplify the radical expression by factoring

out a 4 to obtain $\sqrt{4(b^2 - 30w)}$? Yes ____ No ____

Did you simplify this expression further to get

$2\sqrt{b^2 - 30w}$? Yes ____ No ____

If you answered No to these questions, remember to always simplify radicals whenever possible. Make sure to write your solution as a simplified rational expression.

Now go back and rework the problem using these suggestions.

Find the Vertex, Intercepts and Graph of a Quadratic Function—Problem 17

Find the vertex and the intercepts of $f(x) = -x^2 - 6x - 5$. Then graph the function.

> **Helpful Hint:** When the function is written in $f(x) = ax^2 + bx + c$ form, we can find the vertex using the vertex formula. We can solve for the intercepts using the substitutions $x = 0$ and $f(x) = 0$ to find the unknown coordinates. And, if $a < 0$, the graph is a parabola opening downward.

Do you see that $a = -1$, $b = -6$, and $c = -5$?

Yes ____ No ____

Did you use the vertex formula to discover that the vertex point has an x-coordinate of -3? Yes ____ No ____

If you answered No to these questions, notice that the function is written in $f(x) = ax^2 + bx + c$ form and review

the vertex formula: $x = \dfrac{-b}{2a}$. Substitute the resulting value

for x into the original function to find the y-coordinate of the vertex point.

To find the y-intercept, did you substitute 0 for x into the original function to find the value for y?

Yes ____ No ____

After letting $f(x) = 0$ and substituting the values for a, b, and c into the quadratic formula, did you get the expression

$x = \dfrac{-(-6) \pm \sqrt{(-6)^2 - 4(-1)(-5)}}{2(-1)}$? Yes ____ No ____

If you answered No to these questions, remember that the y-intercept will be an ordered pair in the form $(0, y)$ or in this case, $(0, f(x))$, and the x-intercept will be an ordered pair in the form $(x, 0)$. Be careful when substituting values for a, b, and c into the quadratic formula and remember to evaluate $\sqrt{16}$ as both 4 and -4. Simplify the expression to find the possible x-values.

Since $a < 0$, the parabola will open downward. Plot the vertex, x-intercept, and y-intercept points and connect these points with a curve to find the graph of the function.

If you answered Problem 17 incorrectly, go back and rework the problem using these suggestions.

Chapter 10 The Conic Sections
10.1 The Distance Formula and the Circle

Vocabulary

conic section • distance formula • circle • radius • center

1. A _____ is defined as the set of all points in a plane that are a fixed distance from a point in that plane.

2. The _____ states that the distance between two points (x_1, y_1) and (x_2, y_2) is $d = \sqrt{(x_2 - x_1)^2 + (y_2 - y_1)^2}$.

3. A _____ is a shape formed by slicing a cone with a plane.

4. The point from which the set of points in a circle are a fixed distance from is called the _____.

Example	Student Practice
1. Find the distance between $(3, -4)$ and $(-2, -5)$.	**2.** Find the distance between $(5, 10)$ and $(3, 5)$.

To use the distance formula, we arbitrarily let $(x_1, y_1) = (3, -4)$ and $(x_2, y_2) = (-2, -5)$. Substitute the appropriate values into the formula.

$$d = \sqrt{(x_2 - x_1)^2 + (y_2 - y_1)^2}$$

$$= \sqrt{(-2-3)^2 + \left[-5-(-4)\right]^2}$$

$$= \sqrt{(-5)^2 + (-5+4)^2}$$

$$= \sqrt{(-5)^2 + (-1)^2}$$

$$= \sqrt{25+1}$$

$$= \sqrt{26}$$

Vocabulary Answers: 1. circle 2. distance formula 3. conic section 4. center

Example	Student Practice
3. Find the center and radius of the circle $(x-2)^2+(y-3)^2=25$. Then sketch its graph.	**4.** Find the center and radius of the circle $(x-1)^2+(y-4)^2=16$. Then sketch its graph.

3. Find the center and radius of the circle $(x-2)^2+(y-3)^2=25$. Then sketch its graph.

From the equation of a circle, $(x-h)^2+(y-k)^2=r^2$, we see that $(h,k)=(2,3)$. Thus, the center of the circle is at $(2,3)$. Since $r^2=25$, the radius of the circle is $r=5$.

To graph the circle, start by graphing the center. Then, use the radius to graph the circle. The graph is shown below.

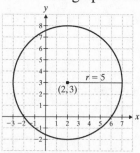

4. Find the center and radius of the circle $(x-1)^2+(y-4)^2=16$. Then sketch its graph.

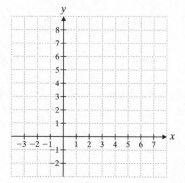

5. Write the equation of the circle with center $(-1,3)$ and radius $\sqrt{5}$. Put your answer in standard form.

We are given that $(h,k)=(-1,3)$ and $r=\sqrt{5}$. Substitute the values into the standard form, $(x-h)^2+(y-k)^2=r^2$.

$$(x-h)^2+(y-k)^2=r^2$$
$$\left[x-(-1)\right]^2+(y-3)^2=\left(\sqrt{5}\right)^2$$
$$(x+1)^2+(y-3)^2=5$$

Be careful of the signs. It is easy to make a sign error in these steps.

6. Write the equation of the circle with center $(14,-5)$ and radius $\sqrt{7}$. Put your answer in standard form.

Example	Student Practice
7. Write the equation of the circle $x^2 + 2x + y^2 + 6y + 6 = 0$ in standard form. Find the radius and center of the circle and sketch its graph.	**8.** Write the equation of the circle $x^2 + 4x - 2 + y^2 - 2y = 2$ in standard form. Find the radius and center of the circle and sketch its graph.

If we multiply out the terms in the standard form of the equation of a circle, we have the following.

$$(x-h)^2 + (y-k)^2 = r^2$$
$$\left(x^2 - 2hx + h^2\right) + \left(y^2 - 2ky + k^2\right) = r^2$$

Comparing this with the given equation, $\left(x^2 + 2x\right) + \left(y^2 + 6y\right) = -6,$ suggests we can complete the squares to put the equations in standard form.

$$x^2 + 2x + \underline{\quad} + y^2 + 6y + \underline{\quad} = -6$$
$$x^2 + 2x + 1 + y^2 + 6y + 9 = -6 + 1 + 9$$
$$x^2 + 2x + 1 + y^2 + 6y + 9 = 4$$
$$(x+1)^2 + (y+3)^2 = 4$$

Thus, the center is at $(-1, -3),$ and the radius is 2. The sketch of the circle is shown below.

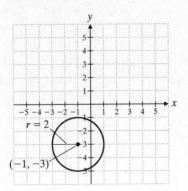

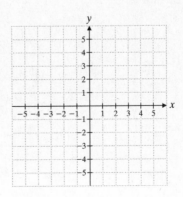

Extra Practice

1. Find the distance between $(-0.5, 8.2)$ and $(3.5, 6.2)$.

2. Write the equation of the circle with center $\left(0, \dfrac{6}{5}\right)$ and radius $\sqrt{13}$. Put your answer in standard form.

3. Find the center and radius of the circle $x^2 + y^2 = 36$. Then sketch its graph.

4. Write the equation of the circle $x^2 + 10x + y^2 - 6y + 29 = 0$ in standard form. Find the radius and center of the circle.

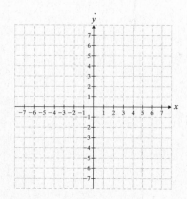

Concept Check

Explain how you would find the values of the unknown coordinate x if the distance between $(-6, 8)$ and $(x, 12)$ is 4.

Name: _____ Date: _____
Instructor: _____ Section: _____

Chapter 10 The Conic Sections
10.2 The Parabola

Vocabulary
parabola • directrix • focus • axis of symmetry • vertex

1. The point at which the parabola crosses the axis of symmetry is the _____.

2. A(n) _____ is defined as the set of points in a plane that are a fixed distance from some fixed line and some fixed point that is not on the line.

3. The fixed point that a parabola is a fixed distance from is called a(n) _____.

Example	Student Practice
1. Graph $y = (x-2)^2$. Identify the vertex and the axis of symmetry.	**2.** Graph $y = (x+1)^2$. Identify the vertex and the axis of symmetry.

Make a table of values. Begin with $x = 2$ in the middle of the table of values because $(2-2)^2 = 0$. That is, when $x = 2$, $y = 0$. Then fill in the x- and y-values above and below $x = 2$.

x	4	3	2	1	0
y	4	1	0	1	4

Plot the points and draw the graph.

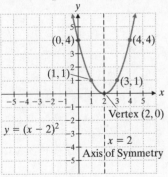

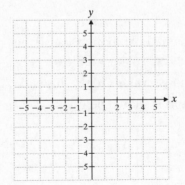

The vertex is $(2,0)$, and the axis of symmetry is the line $x = 2$.

Vocabulary Answers: 1. vertex 2. parabola 3. focus

Example	Student Practice

3. Graph $y = -\frac{1}{2}(x+3)^2 - 1$.

Rewrite the equation in standard form.

$$y = -\frac{1}{2}\left[x - (-3)\right]^2 + (-1)$$

Thus, $a = -\frac{1}{2}$, $h = -3$, and $k = -1$, so it is a vertical parabola. The parabola opens downward since $a < 0$. The vertex is $(-3, -1)$ and the axis of symmetry is the line $x = -3$. The y-intercept is $(0, -5.5)$. Plot a few points on either side of the axis of symmetry.

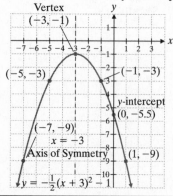

4. Graph $y = -\frac{1}{4}(x+2)^2 - 4$.

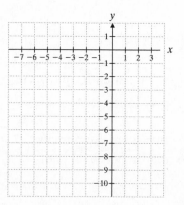

---

5. Graph $x = -2y^2$.

This is a horizontal parabola because the y term is squared. Make a table of values, but choose values for y instead of x. The graph is shown below.

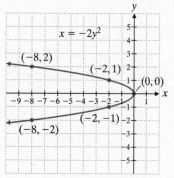

6. Graph $x = -3y^2$.

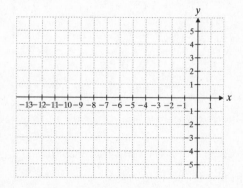

Example	**Student Practice**

7. Place the equation $x = y^2 + 4y + 1$ in standard form. Then graph it.

$x = y^2 + 4y + 1$

$\quad = y^2 + 4y + \left(\dfrac{4}{2}\right)^2 - \left(\dfrac{4}{2}\right)^2 + 1$

$\quad = (y+2)^2 - 3$

Notice that $a = 1$, $k = -2$, and $h = -3$. Since a is positive, the parabola opens to the right. The vertex is $(-3, -2)$ and the axis of symmetry is $y = -2$. Let $y = 0$, to find the x-intercept, $(1, 0)$.

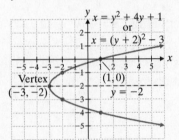

8. Place the equation $x = y^2 + 6y + 7$ in standard form. Then graph it.

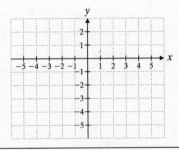

9. Place the equation $y = 2x^2 - 4x - 1$ in standard form. Then graph it.

$y = 2x^2 - 4x - 1$

$\quad = 2\left[x^2 - 2x + (1)^2\right] - 2(1)^2 - 1$

$\quad = 2(x-1)^2 - 3$

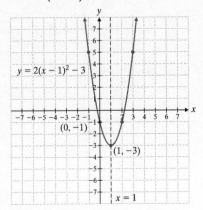

10. Place the equation $y = -2x^2 - 8x - 4$ in standard form. Then graph it.

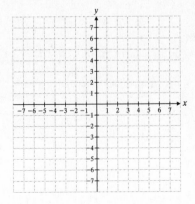

Extra Practice

1. Graph $y = 2(x+4)^2 - 3$ and label the vertex. Find the y-intercept.

2. Graph $y = -2\left(x - \dfrac{3}{2}\right)^2 + 3$ and label the vertex. Find the y-intercept.

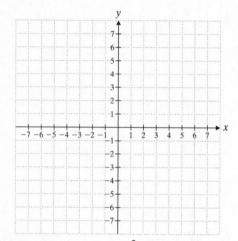

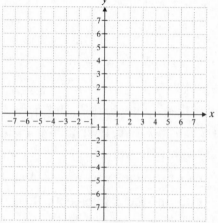

3. Graph $x = 3(y+1)^2 - 3$ and label the vertex. Find the x-intercept.

4. Place the equation $x = -3y^2 + 12y + 6$ in standard form. Determine (a) whether the parabola is horizontal or vertical, (b) the direction it opens, and (c) the vertex.

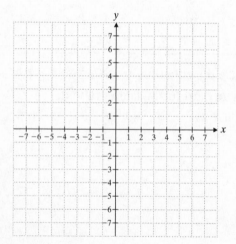

Concept Check

Explain how you can tell which way the parabola opens for these equations: $y = 2x^2$, $y = -2x^2$, $x = 2y^2$, and $x = -2y^2$.

294

Chapter 10 The Conic Sections
10.3 The Ellipse

Vocabulary

ellipse   •   foci   •   vertices   •   center at the origin   •   center at (h, k)

1. The fixed points in an ellipse are called _____.

2. We define a(n) _____ as the set of points in a plane such that for each point in the set, the sum of its distances to two fixed points is constant.

Example	Student Practice
1. Graph $x^2 + 3y^2 = 12$. Label the intercepts.	**2.** Graph $9x^2 + 16y^2 = 144$. Label the intercepts.

Example

Rewrite the equation in standard form.

$$x^2 + 3y^2 = 12$$

$$\frac{x^2}{12} + \frac{3y^2}{12} = \frac{12}{12}$$

$$\frac{x^2}{12} + \frac{y^2}{4} = 1$$

Thus, we have the following:

$a^2 = 12$ so $a = 2\sqrt{3}$

$b^2 = 4$ so $b = 2$

The x-intercepts are $\left(-2\sqrt{3}, 0\right)$ and $\left(2\sqrt{3}, 0\right)$, and the y-intercepts are $(0, 2)$ and $(0, -2)$. We plot these points and draw the ellipse.

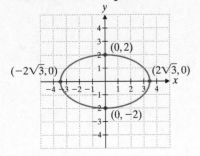

Student Practice

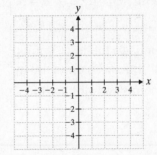

Vocabulary Answers: 1. foci   2. ellipse

Example	Student Practice

3. Graph $\dfrac{(x-5)^2}{9} + \dfrac{(y-6)^2}{4} = 1$.

Notice that this ellipse has the form

$$\dfrac{(x-h)^2}{a^2} + \dfrac{(y-k)^2}{b^2} = 1.$$

The center is (h,k). Note that a and b are not the x-intercepts and y-intercepts now. You'll see that a is the horizontal distance from the center of the ellipse to a point on the ellipse. Similarly, b is the vertical distance. Hence, when the center of the ellipse is not at the origin, the ellipse may not cross either axis.

The center of the ellipse is $(5,6)$, $a = 3$, and $b = 2$. Therefore, we begin at $(5,6)$. We plot points 3 units to the left, 3 units to the right, 2 units up, and 2 units down from $(5,6)$. The points we plot are the vertices of the ellipse.

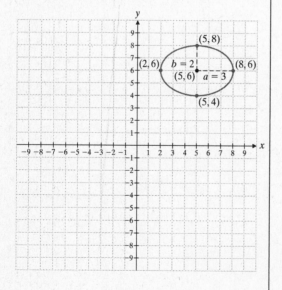

4. Graph $\dfrac{(x-2)^2}{16} + \dfrac{(y-3)^2}{4} = 1$.

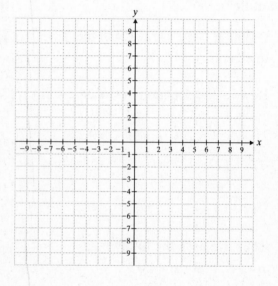

Extra Practice

1. Graph $\dfrac{x^2}{4} + \dfrac{y^2}{36} = 1$. Label the intercepts.

2. Graph $x^2 + 4y^2 = 16$. Label the intercepts.

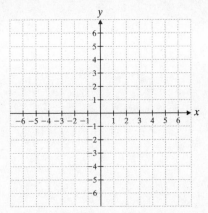

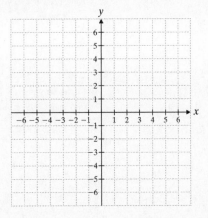

3. Graph $\dfrac{(x+2)^2}{16} + \dfrac{(y-3)^2}{9} = 1$. Label the center.

4. Graph $\dfrac{(x-2)^2}{9} + \dfrac{y^2}{16} = 1$. Label the center.

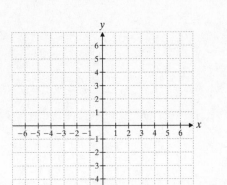

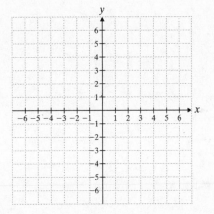

Concept Check

Explain how you would find the four vertices of the ellipse $12x^2 + y^2 - 36 = 0$.

Name: _____   Date: _____

Instructor: _____   Section: _____

Chapter 10 The Conic Sections
10.4 The Hyperbola

Vocabulary
hyperbola • foci • axis • vertices • asymptotes • fundamental rectangle

1. The points where the hyperbola intersects its axis are called the _____.

2. We define a(n) _____ as the set of points in a plane such that for each point in the set, the absolute value of the difference of its distances to two fixed points is constant.

Example	Student Practice
1. Graph $\dfrac{x^2}{25} - \dfrac{y^2}{16} = 1$.	**2.** Graph $\dfrac{x^2}{4} - \dfrac{y^2}{9} = 1$.

The equation has the form $\dfrac{x^2}{a^2} - \dfrac{y^2}{b^2} = 1$, so it is a horizontal hyperbola. $a^2 = 25$, so $a = 5$; $b^2 = 16$, so $b = 4$. Since the hyperbola is horizontal, it has vertices $(a,0)$ and $(-a,0)$ or $(5,0)$ and $(-5,0)$. To draw the asymptotes, we construct a fundamental rectangle with corners at $(5,4)$, $(5,-4)$, $(-5,4)$, and $(-5,-4)$. We draw extended diagonals of the rectangle as the asymptotes. We construct each branch of the curve so that it passes through a vertex and gets closer to the asymptotes as it moves away from the origin.

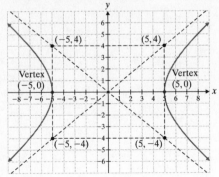

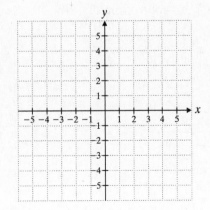

Vocabulary Answers: 1. vertices 2. hyperbola

Example	**Student Practice**
3. Graph $4y^2 - 7x^2 = 28$.	**4.** Graph $3y^2 - 2x^2 = 18$.

3. Graph $4y^2 - 7x^2 = 28$.

To find the vertices and asymptotes, we must rewrite the equation in standard form. Divide each term by 28.

$$4y^2 - 7x^2 = 28$$

$$\frac{4y^2}{28} - \frac{7x^2}{28} = \frac{28}{28}$$

$$\frac{y^2}{7} - \frac{x^2}{4} = 1$$

Thus, we have the standard form of a vertical hyperbola with center at the origin.

Here $b^2 = 7$, so $b = \sqrt{7}$; $a^2 = 4$, so $a = 2$.

The hyperbola has vertices at $\left(0, \sqrt{7}\right)$ and $\left(0, -\sqrt{7}\right)$. The fundamental rectangle has corners at $\left(2, \sqrt{7}\right)$, $\left(2, -\sqrt{7}\right)$, $\left(-2, \sqrt{7}\right)$, and $\left(-2, -\sqrt{7}\right)$.

To aid us in graphing, we measure the distance $\sqrt{7}$ as approximately 2.6.

4. Graph $3y^2 - 2x^2 = 18$.

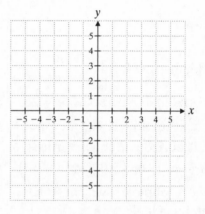

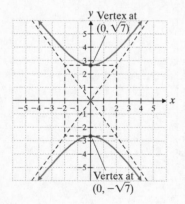

Example	Student Practice

5. Graph $\dfrac{(x-4)^2}{9} - \dfrac{(y-5)^2}{4} = 1.$

6. Graph $\dfrac{(x-1)^2}{4} - \dfrac{(y-3)^2}{16} = 1.$

The center is at $(4,5),$ and the hyperbola is horizontal. We have $a = 3$ and $b = 2,$ so the vertices are $(4 \pm 3, 5),$ or $(7,5)$ and $(1,5).$ We can sketch the hyperbola more readily if we draw a fundamental rectangle. Using $(4,5)$ as the center, we construct a rectangle $2a$ units wide and $2b$ units high. We then draw and extend the diagonals of the rectangle. The extended diagonals are the asymptotes for the branches of the hyperbola.

In this example, since $a = 3$ and $b = 2,$ we draw a rectangle $2a = 6$ units wide and $2b = 4$ units high with a center at $(4,5).$ We draw extended diagonals through the rectangle. From the vertex at $(7,5),$ we draw a branch of the hyperbola opening to the right. From the vertex at $(1,5),$ we draw a branch of the hyperbola opening to the left. The graph of the hyperbola is shown.

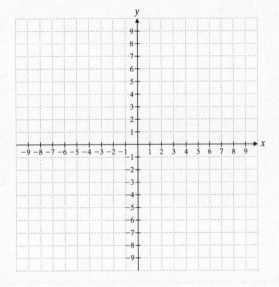

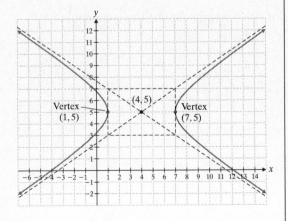

Extra Practice

1. Graph $\dfrac{x^2}{4} - \dfrac{y^2}{36} = 1.$

2. Graph $y^2 - x^2 = 16.$

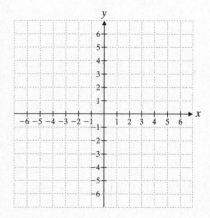

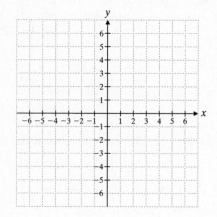

3. Graph $\dfrac{(x+2)^2}{4} - \dfrac{(y+1)^2}{9} = 1.$

4. Graph $\dfrac{(y+1)^2}{9} - \dfrac{(x-1)^2}{9} = 1.$

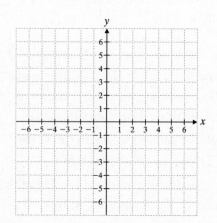

Concept Check

Explain how you would find the equation of one of the asymptotes for the hyperbola $49x^2 - 4y^2 = 196.$

Chapter 10 The Conic Sections
10.5 Nonlinear Systems of Equations

Vocabulary
nonlinear equation   •   nonlinear system of equations

1. A _____ includes at least one nonlinear equation.

2. Any equation that is of second degree or higher is a _____.

Example	Student Practice
1. Solve the following nonlinear system and verify your answer with a sketch.	**2.** Solve the following nonlinear system and verify your answer with a sketch.

Example

1. Solve the following nonlinear system and verify your answer with a sketch.

$$x + y - 1 = 0 \qquad (1)$$
$$y - 1 = x^2 + 2x \quad (2)$$

We will use the substitution method. Solve for y in equation (1).

$$x + y - 1 = 0$$
$$y = -x + 1 \quad (3)$$

Substitute (3) into equation (2). Then solve the resulting quadratic equation.

$$y - 1 = x^2 + 2x$$
$$(-x + 1) - 1 = x^2 + 2x$$
$$-x + 1 - 1 = x^2 + 2x$$
$$0 = x^2 + 3x$$
$$0 = x(x + 3)$$
$$x = 0 \quad \text{or} \quad x = -3$$

Now substitute the values for x in the equation $y = -x + 1$.
Continued on the next page.

Student Practice

2. Solve the following nonlinear system and verify your answer with a sketch.

$$x + y - 5 = 0$$
$$y - 5 = x^2 + 10x$$

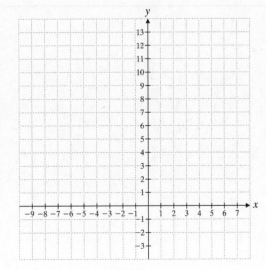

Example	Student Practice

Example

For $x = -3$:

$$y = -(-3) + 1 = +3 + 1 = 4$$

For $x = 0$:

$$y = -(0) + 1 = +1 = 1$$

Thus, the solutions of the system are $(-3, 4)$ and $(0, 1)$.

Sketch the system. Equation (2) describes a parabola. Write it in the following form.

$$y = x^2 + 2x + 1 = (x+1)^2$$

This is a parabola opening upward with its vertex at $(-1, 0)$. Equation (1) can be written as $y = -x + 1$, which is a straight line with slope $= -1$ and y- intercept $(0, 1)$.

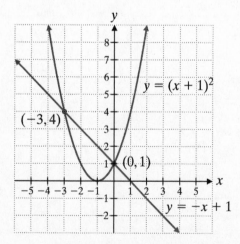

The sketch verifies that the two graphs intersect at $(-3, 4)$ and $(0, 1)$.

Example	Student Practice
3. Solve the following nonlinear system and verify your answer with a sketch.	**4.** Solve the following nonlinear system and verify your answer with a sketch.

Example

3. Solve the following nonlinear system and verify your answer with a sketch.

$$y - 2x = 0 \qquad (1)$$

$$\frac{x^2}{4} + \frac{y^2}{9} = 1 \qquad (2)$$

Solving equation (1) for y yields

$y = 2x$ (3).

Substitute (3) into equation (2).

$$\frac{x^2}{4} + \frac{(2x)^2}{9} = 1$$

$$36\left(\frac{x^2}{4}\right) + 36\left(\frac{(2x)^2}{9}\right) = 36(1)$$

$$9x^2 + 16x^2 = 36$$

$$x = \pm\sqrt{\frac{36}{25}}$$

$$x = \pm1.2$$

For $x = +1.2$: $y = 2(1.2) = 2.4$.

For $x = -1.2$: $y = 2(-1.2) = -2.4$.

We recognize $\frac{x^2}{4} + \frac{y^2}{9} = 1$ as an ellipse

and $y = 2x$ as a straight line. The sketch shows that the points of intersection at $(1.2, 2.4)$ and $(-1.2, -2.4)$ are reasonable.

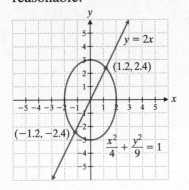

Student Practice

4. Solve the following nonlinear system and verify your answer with a sketch.

$$y + 2x = 0$$

$$x^2 + \frac{y^2}{4} = 1$$

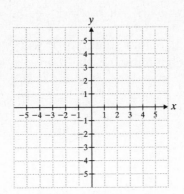

Example	Student Practice
5. Solve the system.	**6.** Solve the system.

5. Solve the system.
$$4x^2 + y^2 = 1 \quad (1)$$
$$x^2 + 4y^2 = 1 \quad (2)$$

Since neither equation is linear, we will use the addition method. Multiply equation (1) by -4 and add to equation (2).

$$-16x^2 - 4y^2 = -4$$
$$\underline{x^2 + 4y^2 = 1}$$
$$-15x^2 = -3$$

$$x^2 = \frac{-3}{-15}$$

$$x^2 = \frac{1}{5}$$

$$x = \pm\sqrt{\frac{1}{5}}$$

If $x = +\sqrt{\dfrac{1}{5}}$, then $x^2 = \dfrac{1}{5}$. Substituting this value into equation (2) gives

$$\frac{1}{5} + 4y^2 = 1$$

$$4y^2 = \frac{4}{5}$$

$$y = \pm\sqrt{\frac{1}{5}}$$

Similarly, if $x = -\sqrt{\dfrac{1}{5}}$, then $y = \pm\sqrt{\dfrac{1}{5}}$. In this case, we have four solutions. If we rationalize each expression, the four

solutions are $\left(\dfrac{\sqrt{5}}{5}, \dfrac{\sqrt{5}}{5}\right)$, $\left(\dfrac{\sqrt{5}}{5}, -\dfrac{\sqrt{5}}{5}\right)$,

$\left(-\dfrac{\sqrt{5}}{5}, \dfrac{\sqrt{5}}{5}\right)$, and $\left(-\dfrac{\sqrt{5}}{5}, -\dfrac{\sqrt{5}}{5}\right)$.

6. Solve the system.
$$9x^2 + y^2 = 1 \quad (1)$$
$$x^2 + 9y^2 = 1 \quad (2)$$

Extra Practice

1. Solve the following nonlinear system and verify your answer with a sketch.

$$y = x^2 - 4$$
$$y = x + 2$$

2. Solve the following nonlinear system by the substitution method.

$$x^2 - 49y^2 - 25 = 0$$
$$x + 7y - 2 = 0$$

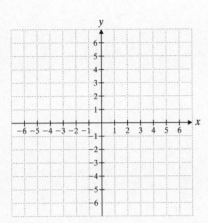

3. Solve the following nonlinear system and verify your answer with a sketch.

$$x^2 + y^2 = 25$$
$$20x^2 - 5y^2 = 100$$

4. Solve the following nonlinear system by the addition method.

$$4x^2 = 3y^2 + 24$$
$$2(x^2 - 15) = -3y^2$$

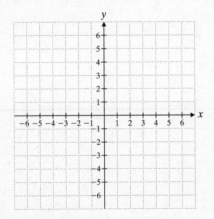

Concept Check

Explain how you would solve the following system.

$$y^2 + 2x^2 = 18$$
$$xy = 4$$

307

MATH COACH

Mastering the skills you need to do well on the test.

Watch the **MATH COACH** videos in MyMathLab®or on You[Tube]™ while you work the problems below. These helpful hints will help you avoid making common errors on test problems.

Rewriting and Graphing the Equation of a Parabola—
Problem 2 Rewrite the equation in standard form. Find the center or vertex, plot at least one other point, identify the conic, and sketch the curve.

$$y^2 - 6y - x + 13 = 0$$

> **Helpful Hint:** If the equation of the parabola has a y^2 term, then it can be written in standard form $x = a(y-k)^2 + h$. The vertex is (h,k), and the graph is a horizontal parabola.

Did you first rewrite the equation as $x = y^2 - 6y + 13$?
Yes _____ No _____
Next, did you complete the square to obtain

$x = (y-3)^2 + 4$? Yes _____ No _____

If you answered No to these questions, remember that when completing the square, we take half of −6 and square it, which results in 9. We must add both a positive and a negative 9 to the right side of the equation. Go back and try to complete these steps again.

Did you see that since $a = 1$ and $a > 0$, the parabola opens to the right? Yes _____ No _____

Did you identify the vertex as $(4,3)$?
Yes _____ No _____

If you answered No to these questions, review the helpful hint and other information about horizontal parabolas. Stop and perform this step again.

In your final answer, be sure to give the equation in standard form, list the vertex, identify the conic, and graph the equation.

If you answered Problem 2 incorrectly, go back and rework the problem using these suggestions.

Identifying and Graphing a Hyperbola with a Center at the Origin—Problem 5

Identify and graph each conic section. Label the center and/or vertex as appropriate. $\dfrac{x^2}{10} - \dfrac{y^2}{9} = 1$

> **Helpful Hint:** The standard form of a hyperbola with the center at the origin and vertices $(-a,0)$ and $(a,0)$ has the equation $\dfrac{x^2}{a^2} - \dfrac{y^2}{b^2} = 1$.

Did you realize that this equation represents a horizontal hyperbola with vertices $(-\sqrt{10},0)$ and $(\sqrt{10},0)$?

Yes _____ No _____
If you answered No, review the rules for the standard form of an equation of a hyperbola with its center at the origin and read the Helpful Hint. Then remember that since $a^2 = 10$ and $a > 0$, $a = \sqrt{10}$. Likewise, if $b^2 = 9$ and $b > 0$, then $b = 3$.
Did you obtain a fundamental rectangle with corners at $(\sqrt{10},3)$, $(\sqrt{10},-3)$, $(-\sqrt{10},3)$, and $(-\sqrt{10},-3)$?
Yes _____ No _____

If you answered No, note that drawing a fundamental rectangle with corners at (a,b), $(a,-b)$, $(-a,b)$, and $(-a,-b)$ can help in creating the graph. The extended diagonals of the rectangle become asymptotes.

In your final answer, remember to identify the conic section, create its graph, and label the center and the vertices.

Now go back and rework the problem using these suggestions.

309

Identifying and Graphing an Ellipse With a Center Not at the Origin—Problem 7

Identify and graph each conic section. Label the center and/or vertex as appropriate. $\dfrac{(x+2)^2}{16}+\dfrac{(y-5)^2}{4}=1$

Helpful Hint: When an ellipse has a center that is not at the origin, the center has the coordinates (h,k) and the equation in standard form is $\dfrac{(x-h)^2}{a^2}+\dfrac{(y-k)^2}{b^2}=1$, where both a and b are greater than zero.

Did you determine that the center of the ellipse is at $(-2,5)$ with $a=4$ and $b=2$?

Yes _____   No _____

If you answered No, remember that the standard form of the equation involves $x-h$ and $y-k$, so you must be careful in determining the signs of h and k.

Did you start at the center and find points a units to the left, a units to the right, b units up, and b units down to plot the points $(-6,5)$, $(2,5)$, $(-2,7)$, and $(-2,3)$?

Yes _____   No _____

If you answered No, remember that to find these four points you need to find the following: $(h-a,k)$, $(h+a,k)$, $(h,k+b)$, and $(h,k-b)$.

Plot all four points and the center and label these on your graph. Then use the four points to make a sketch of the ellipse. Remember to identify the conic as an ellipse in your final answer.

If you answered Problem 7 incorrectly, go back and rework the problem using these suggestions.

Solving a System of Nonlinear Equations—Problem 16

Solve. $\begin{aligned} x^2+2y^2 &= 15 \\ x^2-y^2 &= 6 \end{aligned}$

Helpful Hint: When two equations in a system have the form $ax^2+by^2=c$, where a, b, and c are real numbers, then it may be easiest to solve the system by the addition method.

If you multiplied the second equation by 2 and added the result to the first equation, do you get $3x^2=27$?

Yes _____   No _____

Can you solve this equation for x to get $x=3$ and $x=-3$?

Yes _____   No _____

If you answered No to these questions, remember that when you add the two equations together, the y^2 term adds to 0.

If you substitute $x=3$ into the first equation, do you get the equation $9+2y^2=15$? Yes _____   No _____

Can you solve this equation for y to get $y=\sqrt{3}$ and $y=-\sqrt{3}$? Yes _____   No _____

If you answered No to these questions, try substituting $x=3$ into the equation again and be careful to avoid calculation errors. Remember that you must also perform this same step with $x=-3$. Since x is squared, your results for y should be the same.

Your final answer should have a total of four possible ordered pair solutions to this system.

Now go back and rework this problem using these suggestions.

Chapter 11 Additional Properties of Functions
11.1 Function Notation

Vocabulary
function notation   •   free-fall   •   radius   •   surface area

1. The surface area of a sphere is a function of _____.

2. The approximate distance an object in _____ travels when there is no initial downward velocity is given by the distance function $d(t) = 16t^2$.

Example	**Student Practice**
1. If $g(x) = 5 - 3x$, find the following.	**2.** If $h(x) = 6x - 7$, find the following.
(a) $g(a)$	**(a)** $h(b)$
$g(a) = 5 - 3a$	
(b) $g(a+3)$	
$g(a+3) = 5 - 3(a+3) = 5 - 3a - 9$ $= -4 - 3a$	**(b)** $h(b+6)$
(c) $g(a) + g(3)$	
This requires us to find each addend separately, then add them together.	
$g(a) = 5 - 3a$ $g(3) = 5 - 3(3) = 5 - 9 = -4$	**(c)** $h(b) + h(6)$
$g(a) + g(3) = (5 - 3a) + (-4)$ $= 5 - 3a - 4$ $= 1 - 3a$	
Notice that $g(a+3) \neq g(a) + g(3)$.	

Vocabulary Answers: 1. radius 2. free-fall

Example	Student Practice

3. If $r(x) = \dfrac{4}{x+2}$, find $r(a+3) - r(a)$. Express this result as one fraction.

$$r(a+3) - r(a) = \frac{4}{a+3+2} - \frac{4}{a+2}$$

$$= \frac{4}{a+5} - \frac{4}{a+2}$$

To express this as one fraction, we note that the LCD $= (a+5)(a+2)$.

$$r(a+3) - r(a)$$

$$= \frac{4(a+2)}{(a+5)(a+2)} - \frac{4(a+5)}{(a+2)(a+5)}$$

$$= \frac{4a+8}{(a+5)(a+2)} - \frac{4a+20}{(a+2)(a+5)}$$

$$= \frac{4a-4a+8-20}{(a+5)(a+2)} = \frac{-12}{(a+5)(a+2)}$$

4. If $k(x) = \dfrac{5}{x+6}$, find $k(a+2) - k(a)$. Express this result as one fraction.

---

5. Let $f(x) = 3x - 7$.

Find $\dfrac{f(x+h) - f(x)}{h}$.

First find $f(x+h)$, then subtract $f(x)$.

$$f(x+h) = 3(x+h) - 7 = 3x + 3h - 7$$

$$f(x+h) - f(x)$$
$$= (3x + 3h - 7) - (3x - 7)$$
$$= 3x + 3h - 7 - 3x + 7$$
$$= 3h$$

Therefore, $\dfrac{f(x+h) - f(x)}{h} = \dfrac{3h}{h} = 3$.

6. Let $f(x) = 5x + 8$.

Find $\dfrac{f(x+h) - f(x)}{h}$.

Example	Student Practice
7. The surface area of a sphere is given by $S = 4\pi r^2$ where r is the radius. If we use $\pi = 3.14$ as an approximation, this becomes $S = 4(3.14)r^2$, or $S = 12.56r^2$.	**8.** The surface area of a sphere is given by $S = 4\pi r^2$ where r is the radius. If we use $\pi = 3.14$ as an approximation, this becomes $S = 4(3.14)r^2$, or $S = 12.56r^2$.

(a) Find the surface area of a sphere with a radius of 3 centimeters.

Write surface area as a function of r and solve for $S(r)$ when $r = 3$.

$$S(3) = 12.56(3)^2 = 113.04 \text{ cm}^2$$

(b) Suppose that an error is made and the radius is calculated to be $(3+e)$ centimeters. Find an expression for the surface area as a function of the error e.

$$S(e) = 12.56(3+e)^2$$

$$= 113.04 + 75.36e + 12.56e^2$$

(c) Evaluate the surface area for $r = (3+e)$ cm when $e = 0.2$. Round your answer to the nearest hundredth of a cm. What is the difference in the surface area due to the error in measurement?

An error in measurement was made, so use the function found in part **(b)**.

$$S(0.2)$$

$$= 113.04 + 75.36(0.2) + 12.56(0.2)^2$$

$$= 128.6144$$

$$\approx 128.61 \text{ cm}^2$$

If the radius of 3 cm was incorrectly calculated as 3.2 cm, the surface area would be too large by approximately $128.61 - 113.04 = 15.57 \text{ cm}^2$.

(a) Find the surface area of a sphere with a radius of 6 centimeters.

(b) Suppose that an error is made and the radius is calculated to be $(6+e)$ centimeters. Find an expression for the surface area as a function of the error e.

(c) Evaluate the surface area for $r = (6+e)$ cm when $e = 0.3$. Round your answer to the nearest hundredth of a cm. What is the difference in the surface area due to the error in measurement?

Extra Practice

1. If $g(x) = 5x^2 - 8x + 9$, find $g(a+1)$.

2. If $h(x) = \sqrt{x+2}$, find $h(4a^2 + 6)$.

3. If $s(x) = \dfrac{5}{x+2}$, find $s\left(-\dfrac{4}{3}\right) + s(-3)$.

4. Find $\dfrac{f(x+h) - f(x)}{h}$ for $f(x) = 4x^2$.

Concept Check

Explain how you would find $k(2a-1)$ if $k(x) = \sqrt{3x+1}$.

Chapter 11 Additional Properties of Functions
11.2 General Graphing Procedures for Functions

Vocabulary
function   •   vertical line test   •   vertical shift   •   horizontal shift

1. $f(x)+k$ represents the graph of $f(x)$ with a _____ upwards of k units.

2. A _____ must have no ordered pairs that have the same first coordinates and different second coordinates.

3. The _____ states that if any vertical line intersects the graph of a relation more than once, the relation is not a function.

4. $f(x+h)$ represents the graph of $f(x)$ with a _____ to the left of h units.

Example	**Student Practice**
1. Determine whether the following is the graph of a function. 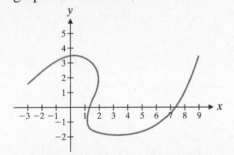	**2.** Determine whether the following is the graph of a function. 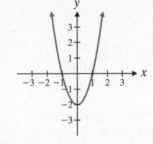
A vertical line intersects the graph more than once, so by the vertical line test, this relation is not a function. 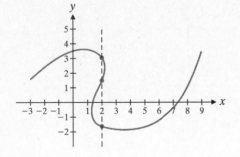	

Vocabulary Answers: 1. vertical shift 2. function 3. vertical line test 4. horizontal shift

Example	Student Practice

3. Graph the functions on one coordinate plane. $f(x) = x^2$ and $h(x) = x^2 + 2$

First we make a table of values for $f(x)$ and for $h(x)$.

x	$f(x) = x^2$
−2	4
−1	1
0	0
1	1
2	4

x	$h(x) = x^2 + 2$
−2	6
−1	3
0	2
1	3
2	6

Graph on the same coordinate plane.

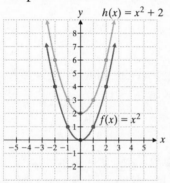

Notice that the graph of $h(x)$ is the graph of $f(x)$ moved 2 units upward.

4. Graph the functions on one coordinate plane. $f(x) = x^2$ and $h(x) = x^2 - 3$

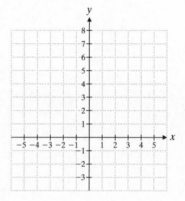

5. Graph the functions on one coordinate plane. $f(x) = |x|$ and $p(x) = |x - 3|$

Graph each function on the same coordinate plane.

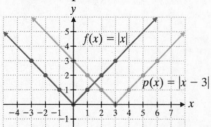

The graph of $p(x)$ is the graph of $f(x)$ shifted 3 units to the right.

6. Graph the functions on one coordinate plane. $f(x) = |x|$ and $p(x) = |x + 3|$

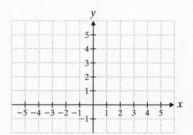

Example	Student Practice

7. Graph the functions on one coordinate plane. $f(x) = x^3$ and $h(x) = (x-3)^3 - 2$

First we make a table of values for $f(x)$ and graph the function.

x	f (x)
−2	−8
−1	−1
0	0
1	1
2	8

Next we recognize that $h(x)$ will have a similar shape, but the curve will be shifted 3 units to the right and 2 units downward. We draw the graph of $h(x)$ using these shifts.

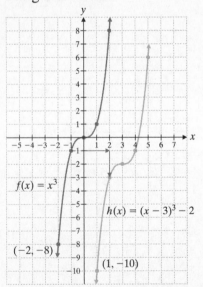

The point $(-2, -8)$ has been shifted 3 units to the right and 2 units down to the point $\left(-2+3, -8+(-2)\right)$ or $(1, -10)$.

The point $(-1, -1)$ is a point on $f(x)$. Use the same reasoning to find the image of $(-1, -1)$ on the graph of $h(x)$. Verify by checking the graphs.

8. Graph the functions on one coordinate plane.

$f(x) = x^3$ and $h(x) = (x+4)^3 - 2$

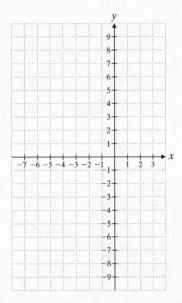

Extra Practice

1. Determine whether the following is the graph of a function.

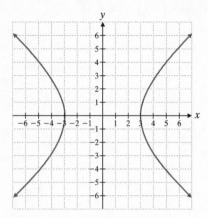

2. Graph the functions on one coordinate plane. $f(x) = x^2$ and $g(x) = (x+3)^2 - 2$

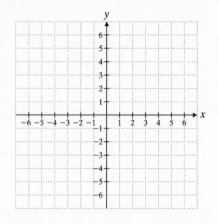

3. Graph the functions on one coordinate plane. $f(x) = x^3$ and $g(x) = (x+2)^3 + 2$

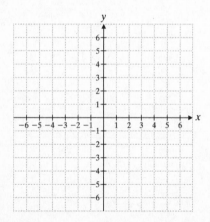

4. Graph the functions on one coordinate plane. $f(x) = \dfrac{3}{x}$ and $g(x) = \dfrac{3}{x-2}$

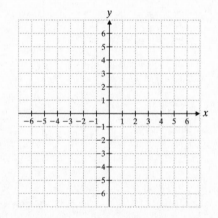

Concept Check

Explain how you can use a vertical line to determine whether a graph represents a function.

Chapter 11 Additional Properties of Functions
11.3 Algebraic Operations on Functions

Vocabulary

sum   •   difference   •   product   •   quotient   •   composition

1. When finding the _____ of a function, we must be careful to avoid division by zero.

2. The _____ of the functions f and g, is defined as follows:
   $(fg)(x) = f(x) \cdot g(x).$

3. The _____ of the functions f and g, denoted $f \circ g$, is defined as follows:
   $(f \circ g)(x) = f[g(x)].$

4. The _____ of the functions f and g, is defined as follows:
   $(f + g)(x) = f(x) + g(x).$

Example	Student Practice
1. Suppose that $f(x) = 3x^2 - 3x + 5$ and $g(x) = 5x - 2$. **(a)** Find $(f + g)(x)$. $$(f + g)(x) = f(x) + g(x)$$ $$= (3x^2 - 3x + 5) + (5x - 2)$$ $$= 3x^2 + 2x + 3$$ **(b)** Evaluate $(f + g)(x)$ when $x = 3$. Write $(f + g)(3)$ and use the formula obtained in **(a)**. $$(f + g)(x) = 3x^2 + 2x + 3$$ $$(f + g)(3) = 3(3)^2 + 2(3) + 3$$ $$= 27 + 6 + 3 = 36$$	**2.** Suppose that $f(x) = 2x^2 + 4x - 7$ and $g(x) = 8x + 1$. **(a)** Find $(f + g)(x)$. **(b)** Evaluate $(f + g)(x)$ when $x = 4$.

Vocabulary Answers: 1. quotient   2. composition   3. product   4. sum

Example	Student Practice
3. Given $f(x) = x^2 - 5x + 6$ and $g(x) = 2x - 1$, find the following. **(a)** $(fg)(x)$ $$(fg)(x) = f(x) \cdot g(x)$$ $$= (x^2 - 5x + 6)(2x - 1)$$ $$= 2x^3 - 11x^2 + 17x - 6$$ **(b)** $(fg)(-4)$ Use the formula obtained in **(a)**. $$(fg)(x) = 2x^3 - 11x^2 + 17x - 6$$ $$(fg)(-4)$$ $$= 2(-4)^3 - 11(-4)^2 + 17(-4) - 6$$ $$= -128 - 176 - 68 - 6 = -378$$	**4.** Given $f(x) = x^2 - 3x + 8$ and $g(x) = 4x - 3$, find the following. **(a)** $(fg)(x)$ **(b)** $(fg)(-2)$
5. Given $f(x) = 3x + 1$, $g(x) = 2x - 1$, and $h(x) = 9x^2 + 6x + 1$, find the following. **(a)** $\left(\dfrac{f}{g}\right)(x)$ $$\left(\dfrac{f}{g}\right)(x) = \dfrac{3x+1}{2x-1}$$ The denominator of the quotient can never be zero. Since $2x - 1 \neq 0$, we know that $x \neq \dfrac{1}{2}$. **(b)** $\left(\dfrac{f}{h}\right)(x)$ $$\left(\dfrac{f}{h}\right)(x) = \dfrac{3x+1}{9x^2+6x+1}$$ $$= \dfrac{3x+1}{(3x+1)(3x+1)} = \dfrac{1}{3x+1}$$ Since $3x + 1 \neq 0$, we know $x \neq -\dfrac{1}{3}$.	**6.** Given $f(x) = 2x + 3$, $g(x) = x - 4$, and $h(x) = 4x^2 + 12x + 9$, find the following. **(a)** $\left(\dfrac{f}{g}\right)(x)$ **(b)** $\left(\dfrac{f}{h}\right)(x)$

320

Example	Student Practice
7. Given $f(x) = 3x - 2$ and $g(x) = 2x + 5$, find $f[g(x)]$.	**8.** Given $f(x) = 5x + 3$ and $g(x) = 3x - 4$, find $f[g(x)]$.

First substitute $g(x) = 2x + 5$. Then apply the formula for $f(x)$. Remove the parentheses and simplify.

$$f[g(x)] = f(2x + 5)$$
$$= 3(2x + 5) - 2$$
$$= 6x + 15 - 2$$
$$= 6x + 13$$

---

9. Given $f(x) = \sqrt{x - 4}$ and $g(x) = 3x + 1$, find the following.

(a) $f[g(x)]$

Substitute $g(x) = 3x + 1$. Then apply the formula for $f(x)$.

$$f[g(x)] = f(3x + 1)$$
$$= \sqrt{(3x + 1) - 4}$$
$$= \sqrt{3x + 1 - 4}$$
$$= \sqrt{3x - 3}$$

(b) $g[f(x)]$

$$g[f(x)] = g(\sqrt{x - 4})$$
$$= 3(\sqrt{x - 4}) + 1$$
$$= 3\sqrt{x - 4} + 1$$

We note that $g[f(x)] \neq f[g(x)]$.

10. Given $f(x) = \sqrt{x + 6}$ and $g(x) = 4x - 1$, find the following.

(a) $f[g(x)]$

(b) $g[f(x)]$

Example	Student Practice

11. Given $f(x) = 2x$ and $g(x) = \dfrac{1}{3x-4}$, $x \neq \dfrac{4}{3}$, find the following.

(a) $(f \circ g)(x)$

$$(f \circ g)(x) = f\left[g(x)\right] = f\left(\dfrac{1}{3x-4}\right)$$
$$= 2\left(\dfrac{1}{3x-4}\right) = \dfrac{2}{3x-4}$$

(b) $(f \circ g)(2)$

$$(f \circ g)(2) = \dfrac{2}{3(2)-4} = \dfrac{2}{6-4} = \dfrac{2}{2} = 1$$

12. Given $f(x) = 4x$ and $g(x) = \dfrac{1}{2x-5}$, $x \neq \dfrac{5}{2}$, find the following.

(a) $(f \circ g)(x)$

(b) $(f \circ g)\left(\dfrac{1}{2}\right)$

Extra Practice

1. Given $f(x) = 1.6x^3 - 2.7x$ and $g(x) = 4.6x^2 + 7.6$, find the following.

(a) $(f+g)(x)$
(b) $(f-g)(x)$
(c) $(f+g)(3)$
(d) $(f-g)(-2)$

2. Given $f(x) = x^2 - 8x + 16$ and $g(x) = x - 4$, find the following.

(a) $(fg)(x)$
(b) $\left(\dfrac{f}{g}\right)(x)$
(c) $(fg)(3)$
(d) $\left(\dfrac{f}{g}\right)(-3)$

3. Given $f(x) = x - 5$ and $g(x) = 6 - 2x$, find $f\left[g(x)\right]$.

4. Given $f(x) = \left|\dfrac{4}{3}x + 1\right|$ and $g(x) = -3x - 2$, find $f\left[g(x)\right]$.

Concept Check

If $f(x) = 5 - 2x^2$ and $g(x) = 3x^2 - 5x + 1$, explain how you would find $(f-g)(-4)$.

Name: _____   Date: _____

Instructor: _____   Section: _____

Chapter 11 Additional Properties of Functions
11.4 Inverse of a Function

Vocabulary

inverse function f •   one-to-one function   •   horizontal line test

1. The _____ states that if any horizontal line intersects the graph of a function more than once, the function is not one-to-one.

2. We call a function f^{-1} that reverses the domain and range of a function f the

   _____.

3. A(n) _____ is a function in which no ordered pairs have the same second coordinate.

Example	Student Practice
1. Indicate whether the following functions are one-to-one.	**2.** Indicate whether the following functions are one-to-one.
(a) $M = \{(1,3),(2,7),(5,8),(6,12)\}$	**(a)** $L = \{(1,6),(4,3),(7,6),(9,16)\}$
M is a function because no ordered pairs have the same first coordinate. M is also a one-to-one function because no ordered pairs have the same second coordinate.	
(b) $P = \{(1,4),(2,9),(3,4),(4,18)\}$	**(b)** $N = \{(2,6),(3,4),(6,9),(7,11)\}$
P is a function, but it is not one-to-one because the ordered pairs $(1,4)$ and $(3,4)$ have the same second coordinate.	

Vocabulary Answers: 1. horizontal line test 2. inverse function f 3. one-to-one function

Example	Student Practice

Example

3. Determine whether the functions graphed are one-to-one functions.

(a)

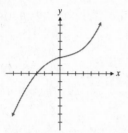

This graph represents a one-to-one function. Horizontal lines cross the graph at most once.

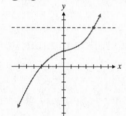

(b)

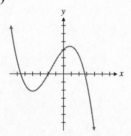

This graph does not represent a one-to-one function. A horizontal line exists that crosses the graph more than once.

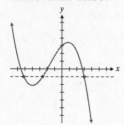

Student Practice

4. Determine whether the functions graphed are one-to-one functions.

(a)

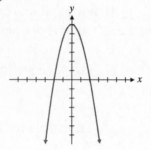

(b)

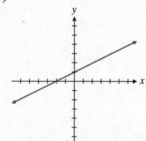

Example	Student Practice
5. Determine the inverse function of $F = \{(6,1),(12,2),(13,5),(14,6)\}$. Since we have a list of ordered pairs, we interchange the coordinates of each ordered pair. The inverse function of F is as follows. $F^{-1} = \{(1,6),(2,12),(5,13),(6,14)\}$	**6.** Determine the inverse function of $Q = \{(2,4),(6,2),(9,7),(11,3)\}$.

7. Find the inverse of $f(x) = 7x - 4$.

Replace $f(x)$ with y.

$y = 7x - 4$

Interchange the variables x and y.

$x = 7y - 4$

Solve for y in terms of x.

$x = 7y - 4$
$x + 4 = 7y$
$\dfrac{x+4}{7} = y$

Replace y with $f^{-1}(x)$.

$f^{-1}(x) = \dfrac{x+4}{7}$

8. Find the inverse of $f(x) = 5x + 8$.

Example	Student Practice
9. Find the inverse function of $f(x) = \frac{9}{5}x + 32$, which converts Celsius temperature (x) into equivalent Fahrenheit temperature. $y = \frac{9}{5}x + 32$ $x = \frac{9}{5}y + 32$ $5x = 9y + 160$ $\frac{5x - 160}{9} = y$ $f^{-1}(x) = \frac{5x - 160}{9}$ Our inverse function $f^{-1}(x)$ will now convert Fahrenheit temperature to Celsius temperature.	**10.** Find the inverse function of $f(x) = 150 + 3(x - 25)$, which gives the cost of a rental truck if the company charges a base rate of \$150 plus \$3 for every mile traveled over 25 miles. Here x is the total number of miles traveled in the rental truck.
11. If $f(x) = 3x - 2$, find $f^{-1}(x)$. Graph f and f^{-1} on the same set of axes. Draw the line $y = x$ as a dashed line for reference. Following the procedure to find $f^{-1}(x)$ yields $f^{-1}(x) = \frac{x+2}{3}$. Now we graph each line. 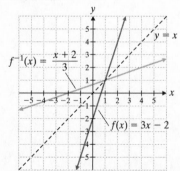 We see that the graphs of f and f^{-1} are symmetric about the line $y = x$.	**12.** If $f(x) = 4x + 1$, find $f^{-1}(x)$. Graph f and f^{-1} on the same set of axes. Draw the line $y = x$ as a dashed line for reference. 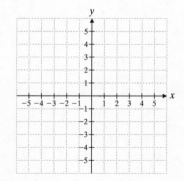

Extra Practice

1. Indicate whether the following function is one-to-one.

   $$B = \{(7,9),(9,7),(-7,-9),(-9,-7)\}$$

2. Find the inverse of $f(x) = x - 4$.

3. Find the inverse of $f(x) = \dfrac{5}{3x - 4}$.

4. Find the inverse of $g(x) = -3x - 4$. Graph the function and its inverse on the same set of axes. Draw the line $y = x$ as a dashed line for reference.

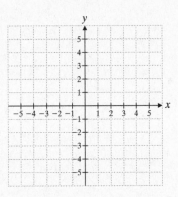

Concept Check

If $f(x) = \dfrac{x - 5}{3}$, explain how you would find the inverse function.

MATH COACH

Mastering the skills you need to do well on the test.

Watch the **MATH COACH** videos in MyMathLab®or on You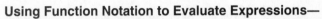Tube™ while you work the problems below. These helpful hints will help you avoid making common errors on test problems.

Using Function Notation to Evaluate Expressions—

Problem 5 For the function $f(x) = 3x^2 - 2x + 4$, find $f(a+1)$.

> **Helpful Hint:** The key idea is to replace every x with the expression $a+1$ and then simplify the result. Use parentheses around the substitutions to avoid calculation errors.

Did you substitute $a+1$ into the function to get

$3(a+1)^2 - 2(a+1) + 4$?

Yes _____ No _____

Did you evaluate $(a+1)^2$ to get $a^2 + 2a + 1$?

Yes _____ No _____

If you answered No to these questions, remember to replace every x with $a+1$. Use parentheses to avoid calculation errors. Note that when you square a binomial, you must be sure to write down all the terms.

Next, did you simplify further to obtain

$3a^2 + 6a + 3 - 2a - 2 + 4$? Yes _____ No _____

If you answered No, remember to multiply all three terms of $a^2 + 2a + 1$ by 3. Multiply both terms of $a+1$ by -2.

In your final step, you can collect like terms to write your answer in simplest form.

If you answered Problem 5 incorrectly, go back and rework the problem using these suggestions.

Graphing a Function with a Horizontal and Vertical Shift—Problem 10
Graph each pair of functions on one coordinate plane. $f(x) = x^2$

$$g(x) = (x-1)^2 + 3$$

> **Helpful Hint:** First graph $f(x)$. The graph of $f(x-h) + k$ is the graph of $f(x)$ shifted h units to the right and k units upward (assuming that $h > 0$ and $k > 0$).

Can you determine that the graph of $f(x)$ passes through the points $(0,0)$, $(1,1)$, $(-1,1)$, $(2,4)$, and $(-2,4)$?

Yes _____ No _____

If you answered No, try building a table of values in which you replace x with 0, 1, -1, 2, and -2 and find the corresponding values of $f(x)$. Plot those points and connect the points with a curve to form the graph of $f(x) = x^2$.

Do you see that the function $g(x) = (x-1)^2 + 3$ has the values of $h = 1$ and $k = 3$ when you apply the Helpful Hint?

Yes _____ No _____

Did you find that the graph of $g(x)$ is the graph of $f(x)$ shifted one unit to the right and 3 units up? Yes _____ No _____

If you answered No to these questions, reread the Helpful Hint carefully. Notice that the values of h and k are both greater than zero.

Finish by graphing $g(x)$ on the same coordinate plane as your graph of $f(x)$.

Now go back and rework the problem using these suggestions.

329

Finding the Composition of Two Functions—Problem 14(a)

If $f(x)=\dfrac{1}{2}x-3$ and $g(x)=4x+5$, find $(f\circ g)(x)$.

> **Helpful Hint:** First rewrite $(f\circ g)(x)$ as $f\big[g(x)\big]$. Most students find this expression more logical. Then substitute $g(x)$ for the value of x in $f(x)$.

First, did you rewrite the problem as

$$f\big[g(x)\big]=\dfrac{1}{2}(4x+5)-3?$$

Yes _____ No _____

If you answered No, substitute the expression for $g(x)$ for the value of x in the expression for $f(x)$.

Next did you simplify the resulting expression to

$$f\big[g(x)\big]=2x+\dfrac{5}{2}-3?$$

Yes _____ No _____

If you answered No, remember that

$\dfrac{1}{2}(4x)=2x$ and $\dfrac{1}{2}(5)=\dfrac{5}{2}$. As your final step, combine like terms to write your expression in simplest form.

If you answered Problem 14(a) incorrectly, go back and rework the problem using these suggestions.

Finding and Graphing the Inverse of a Function—Problem 18

Given $f(x)=-3x+2$, find f^{-1}. Graph f and its inverse f^{-1} on one coordinate plane. Graph $y=x$ as a dashed line for reference.

> **Helpful Hint:** Use the following four steps to find the inverse of a function:
> 1. Replace $f(x)$ with y.
> 2. Interchange x and y.
> 3. Solve for y in terms of x.
> 4. Replace y with $f^{-1}(x)$.

Did you substitute y for $f(x)$ to get $y=-3x+2$ and then interchange x and y to get $x=-3y+2$?

Yes _____ No _____

If you answered No, review the first two steps in the Helpful Hint and perform these steps again.

Did you solve the equation for y to get $y=-\dfrac{1}{3}x+\dfrac{2}{3}$?

Yes _____ No _____

If you answered No, remember to add $3y$ to each side. Next add $-x$ to each side and then divide each side of the equation by 3. As your last step in finding the inverse, replace y with $f^{-1}(x)$.

Remember to graph $f(x)$ and $f^{-1}(x)$ on the same coordinate plane. Add the graph of $y=x$ as a dashed line for reference.

Now go back and rework the problem using these suggestions.

Name: _____   Date: _____
Instructor: _____   Section: _____

Chapter 12 Logarithmic and Exponential Functions
12.1 The Exponential Function

Vocabulary

exponential function • e • asymptote • property of exponential equations
compound interest • radioactive decay • base

1. The _____ says that if $b^x = b^y$, then $x = y$ for $b > 0$ and $b \neq 1$.

2. The function $f(x) = b^x$, where $b > 0$, $b \neq 1$, and x is a real number, is called a(n) _____.

3. For every exponential function, the x-axis is a(n) _____.

Example	**Student Practice**
1. Graph $f(x) = 2^x$.	**2.** Graph $f(x) = 5^x$.

Make a table of values for x and $f(x)$.

$$f(-1) = 2^{-1} = \frac{1}{2}, \ f(0) = 2^0 = 1, \ f(1) = 2^1 = 2$$

Continue to evaluate $f(x)$ at different integers to make a table of values for x and $f(x)$. Using these points, draw the graph.

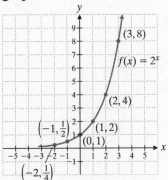

Notice how the curve comes very close to the x-axis but never touches it. The x-axis is an asymptote for every exponential function.

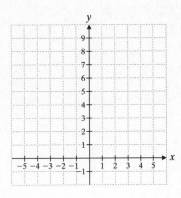

Vocabulary Answers: 1. property of exponential equations 2. exponential function 3. asymptote

Example	Student Practice

3. Graph $f(x) = \left(\dfrac{1}{2}\right)^x$.

Rewrite the function as follows.

$$f(x) = \left(\frac{1}{2}\right)^x = \left(2^{-1}\right)^x = 2^{-x}$$

Evaluate the function for a few values of x. Using these points, draw the graph.

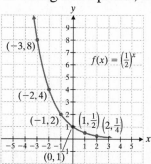

Note that as x increases, $f(x)$ decreases.

4. Graph $f(x) = \left(\dfrac{2}{3}\right)^x$.

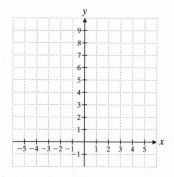

---

5. Graph $f(x) = e^x$.

The letter e is like the number π. It is an irrational number. If you do not have a scientific calculator, approximate the value for e to use the number in calculations, $e \approx 2.7183$. Otherwise, use a scientific calculator to produce a table of values for this function. Then, draw the graph.

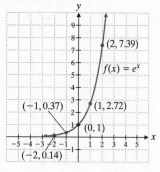

6. Graph $f(x) = e^{1+x}$.

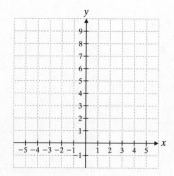

332

Example	Student Practice
7. Solve. $2^x = \dfrac{1}{16}$	**8.** Solve. $3^x = 27$

To use the property of exponential equations, we must have the same base on both sides of the equation. Rewrite 16 as 2^4.

$$2^x = \frac{1}{16}$$

$$2^x = \frac{1}{2^4}$$

$$2^x = 2^{-4}$$

Recall the property of exponential equations, which states that if $b^x = b^y$, then $x = y$ for $b > 0$ and $b \neq 1$. Since $2^x = 2^{-4}$, $x = -4$.

---

9. If we invest $8000 in a fund that pays 15% annual interest compounded monthly, how much will we have in 6 years?

10. If you invest $5500 in a fund that pays 18% annual interest compounded quarterly, how much will you have in 10 years?

In this situation, $P = 8000$, $r = 15\% = 0.15$, $t = 6$, and $n = 12$ because interest is compounded monthly, or 12 times a year.

$$A = 8000\left(1 + \frac{0.15}{12}\right)^{(12)(6)}$$

$$= 8000(1 + 0.125)^{72}$$

$$= 8000(2.445920268)$$

$$\approx 19{,}567.36$$

Example	Student Practice
11. The radioactive decay of the element americium 241 can be described by the equation $A = Ce^{-0.001608t}$, where C is the original amount of the element in the sample, A is the amount of the element remaining after t years, and $k = -0.0016008$, the decay constant for americium 241. If 10 milligrams (mg) of americium 241 is sealed in a laboratory container today, how much will theoretically be present in 2000 years? Round your answer to the nearest hundredth.	**12.** The radioactive decay of the element cobalt 60 can be described by the equation $A = Ce^{-0.131527t}$, where C is the original amount of the element in the sample, A is the amount of the element remaining after t years, and $k \approx -0.131527$, the decay constant for cobalt 60. If 5 milligrams (mg) of cobalt 60 is sealed in a laboratory container today, how much will theoretically be present in 10 years? Round your answer to the nearest hundredth.

Substitute $C = 10$ and $t = 2000$ into the equation and simplify using a calculator.

$$A = Ce^{-0.0016008t} = 10e^{-0.0016008(2000)}$$

$$\approx 10(0.040697)$$

$$\approx 0.41 \text{ mg}$$

Extra Practice

1. Graph $f(x) = 3^{-x}$.

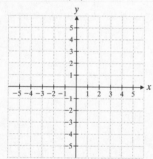

2. Graph $f(x) = 2^{x+4}$.

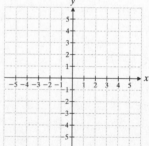

3. Solve for x. $5^{x+3} = 25$

4. Anton is investing $4000 at an annual rate of 4.6% compounded annually. How much money will Anton have after 5 years? Round your answer to the nearest cent.

Concept Check

Explain how you would solve $4^{-x} = \dfrac{1}{64}$.

Chapter 12 Logarithmic and Exponential Functions
12.2 The Logarithmic Function

Vocabulary
logarithmic function   •   logarithm   •   base   •   power   •   inverse

1. The logarithmic function $y = \log_b x$ is the _____ of the exponential function $x = b^y$.

2. A(n) _____ is an exponent.

3. In the function $y = \log_b x$, b is called the _____.

Example	Student Practice
1. Write in logarithmic form. $81 = 3^4$ Use the fact that $x = b^y$ is equivalent to $\log_b x = y$. Here, $x = 81$, $b = 3$, and $y = 4$. So the logarithmic form of $81 = 3^4$ is $4 = \log_3 81$.	**2.** Write in logarithmic form. $\dfrac{1}{216} = 6^{-3}$
3. Write in exponential form. $-4 = \log_{10}\left(\dfrac{1}{10,000}\right)$ Use the fact that $x = b^y$ is equivalent to $\log_b x = y$. Here, $y = -4$, $b = 10$, and $x = \dfrac{1}{10,000}$. So the exponential form of $-4 = \log_{10}\left(\dfrac{1}{10,000}\right)$ is $\dfrac{1}{10,000} = 10^{-4}$.	**4.** Write in exponential form. $5 = \log_3 243$

Vocabulary Answers: 1. inverse 2. logarithm 3. base

Example	Student Practice
5. Solve for the variable.	**6.** Solve for the variable
(a) $\log_5 x = -3$	**(a)** $\log_2 x = -8$
Convert the logarithmic equation to an equivalent exponential equation and solve for x.	
$5^{-3} = x$	
$\dfrac{1}{5^3} = x$	
$\dfrac{1}{125} = x$	
(b) $\log_a 16 = 4$	**(b)** $\log_b 64 = 3$
$a^4 = 16$	
$a^4 = 2^4$	
$a = 2$	
7. Evaluate. $\log_3 81$	**8.** Evaluate. $\log_2 64$
This is asking, "To what power must we raise 3 to get 81?" Write an equivalent exponential expression using x as the unknown power.	
$\log_3 81 = x$	
$81 = 3^x$	
Write 81 as 3^4 and solve for x.	
$3^4 = 3^x$	
$x = 4$	
Thus, $\log_3 81 = 4$.	

336

Example	Student Practice

9. Graph $y = \log_2 x$ and $y = 2^x$ on the same set of axes.

Make a table of values (ordered pairs) for each equation.

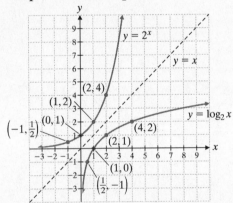

x	y
-1	$\frac{1}{2}$
0	1
1	2
2	4

x	y
$\frac{1}{2}$	-1
1	0
2	1
4	2

Coordinates of ordered pairs are reversed

Graph the ordered pairs.

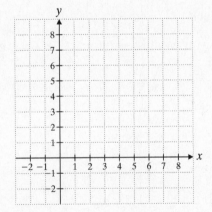

Note that $y = \log_2 x$ is the inverse of $y = 2^x$ because the ordered pairs (x, y) are reversed. The sketch of the two equations shows that they are inverses.

Recall that in function notation, f^{-1} means the inverse function of f. Thus, if we write $f(x) = \log_2 x$, then $f^{-1}(x) = 2^x$.

10. Graph $y = \log_4 x$ and $y = 4^x$ on the same set of axes.

337

Extra Practice

1. Write in logarithmic form. $0.0001 = 10^{-4}$

2. Write in exponential form. $-6 = \log_e x$

3. Evaluate. $\log_{10} \sqrt{10}$

4. Graph. $\log_2 x = y$

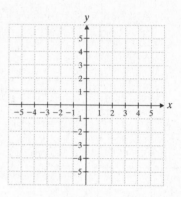

Concept Check

Explain how you would solve for x. $\quad -\dfrac{1}{2} = \log_e x$

Chapter 12 Logarithmic and Exponential Functions
12.3 Properties of Logarithms

Vocabulary
logarithms • Logarithm of a Product Property • Logarithm of a Quotient Property
Logarithm of a Number Raised to a Power Property

1. The _____ states that for any positive real numbers M and N and any positive base $b \neq 1$, $\log_b MN = \log_b M + \log_b N$.

2. The _____ is similar to the logarithm of a product property, except that it involves two expression that are divided.

3. _____ are used to reduce complex expressions to addition and subtraction.

Example	**Student Practice**
1. Write $\log_3 XZ$ as a sum of logarithms.	**2.** Write $\log_5 BC$ as a sum of logarithms.
Use the logarithm of a product property, $\log_b MN = \log_b M + \log_b N$, to rewrite the logarithm.	
$b = 3$, $M = X$, and $N = Z$	
$\log_3 XZ = \log_3 X + \log_3 Z$	
3. Write $\log_3 16 + \log_3 x + \log_3 y$ as a single logarithm.	**4.** Write $\log_2 x + \log_2 20 + \log_2 z$ as a single logarithm.
If we extend our rule, we have $\log_b MNP = \log_b M + \log_b N + \log_b P$.	
Use this to rewrite the logarithm.	
$b = 3$, $M = 16$, $M = x$, and $P = y$	
$\log_3 16 + \log_3 x + \log_3 y = \log_3 16xy$	

Vocabulary Answers: 1. logarithm of a product property 2. logarithm of a quotient property 3. logarithms

Example	Student Practice
5. Write $\log_3\left(\dfrac{29}{7}\right)$ as the difference of two logarithms.	**6.** Write $\log_{10}\left(\dfrac{19}{5}\right)$ as the difference of two logarithms.

5. Write $\log_3\left(\dfrac{29}{7}\right)$ as the difference of two logarithms.

To rewrite the logarithm, use the logarithm of a quotient property,

$$\log_b\left(\frac{M}{N}\right) = \log_b M - \log_b N.$$

$$\log_3\left(\frac{29}{7}\right) = \log_3 29 - \log_3 7$$

6. Write $\log_{10}\left(\dfrac{19}{5}\right)$ as the difference of two logarithms.

---

7. Express $\log_b 36 - \log_b 9$ as a single logarithm.

Use the reverse of the property

$$\log_b\left(\frac{M}{N}\right) = \log_b M - \log_b N \text{ to write}$$

the difference as a single logarithm.

$$\log_b 36 - \log_b 9 = \log_b\left(\frac{36}{9}\right) = \log_b 4$$

8. Express $\log_b 6 - \log_b 24$ as a single logarithm.

---

9. Write $\dfrac{1}{3}\log_b x + 2\log_b w - 3\log_b z$ as a single logarithm.

Use the property $\log_b M^p = p\log_b M$ to eliminate the coefficients of the logarithmic terms. Then, combine the sum of the logarithms and then the difference to obtain a single logarithm.

$$\frac{1}{3}\log_b x + 2\log_b w - 3\log_b z$$

$$= \log_b x^{1/3} + \log_b w^2 - \log_b z^3$$

$$= \log_b x^{1/3}w^2 - \log_b z^3$$

$$= \log_b\left(\frac{x^{1/3}w^2}{z^3}\right)$$

10. Write $5\log_b y - \dfrac{1}{4}\log_b w + 3\log_b z$ as a single logarithm.

Example	Student Practice
11. Write $\log_b\left(\dfrac{x^4 y^3}{z^2}\right)$ as a sum or difference of logarithms.	**12.** Write $\log_b\left(\dfrac{w^6 x^5}{y^3}\right)$ as a sum or difference of logarithms.

Use the logarithm of a quotient property to rewrite the logarithm as a difference.

$$\log_b\left(\frac{x^4 y^3}{z^2}\right) = \log_b x^4 y^3 - \log_b z^2$$

Now, use the logarithm of a product property.

$$\log_b x^4 y^3 - \log_b z^2$$
$$= \log_b x^4 + \log_b y^3 - \log_b z^2$$

Finally, use the logarithm of a number raised to a power property to eliminate the exponents and get the following result.

$$4\log_b x + 3\log_b y - 2\log_b z$$

Example	Student Practice
13. Evaluate.	**14.** Evaluate.
(a) Evaluate $\log_7 7$.	**(a)** Evaluate $\log_{20} 20$.
Since $\log_b b = 1$, $\log_7 7 = 1$.	
(b) Evaluate $\log_5 1$.	**(b)** Evaluate $\log_7 1$.
Since $\log_b 1 = 0$, $\log_5 1 = 0$.	
(c) Find x if $\log_3 x = \log_3 17$.	**(c)** Find x if $\log_5 x = \log_5 12$.
If $\log_b x = \log_b y$, then $x = y$.	
Thus, since $\log_3 x = \log_3 17$, $x = 17$.	

Example	Student Practice
15. Find x if $2\log_7 3 - 4\log_7 2 = \log_7 x$.	**16.** Find x if $2\log_{11} 7 - 3\log_{11} 5 = \log_{11} x$.

Use the logarithm of a number raised to a power property to eliminate the coefficients of the logarithmic terms.

$$\log_7 3^2 - \log_7 2^4 = \log_7 x$$
$$\log_7 9 - \log_7 16 = \log_7 x$$

Then, use the logarithm of a quotient property and then the property that if $\log_b x = \log_b y$, then $x = y$, to find x.

$$\log_7\left(\frac{9}{16}\right) = \log_7 x$$

$$\frac{9}{16} = x$$

Extra Practice

1. Write as a sum or difference of logarithms. $\log_7 x^3 y z^2$

2. Write as a single logarithm. $\frac{1}{3}\log_a 6 + 2\log_a 6 - 5\log_a x$

3. Use the properties of logarithms to simplify the following. $\frac{1}{5}\log_3 3 - \log_{11} 1$

4. Find x if $\log_4 x + \log_4 2 = 3$.

Concept Check

Explain how you would simplify $\log_{10}(0.001)$.

Chapter 12 Logarithmic and Exponential Functions
12.4 Common Logarithms, Natural Logarithms, and Change of Base Logarithms

Vocabulary
common logarithm   •   antilogarithm   •   natural logarithm   •   change of base formula

1. Base 10 logarithms are called _____ and are usually written with no subscripts.

2. The _____ is $\log_b x = \dfrac{\log_a x}{\log_a b}$, where a, b, and $x > 0$, $a \neq 1$, and $b \neq 1$.

3. Logarithms with base e are known as _____ and are usually written $\ln x$.

Example	Student Practice
1. On a scientific calculator or a graphing calculator, find a decimal approximation for each of the following.	**2.** On a scientific calculator or a graphing calculator, find a decimal approximation for each of the following.
(a) $\log 7.32$	**(a)** $\log 12.67$
Enter the number 7.32 on a calculator and then press the log key.	
$\log 7.32 \approx 0.864511081$	
(b) $\log 73.2$	**(b)** $\log 126.7$
$\log 73.2 \approx 1.864511081$	
3. Find an approximate value for x if $\log x = 4.326$.	**4.** Find an approximate value for x if $\log x = 7.438$.
Find the antilogarithm. We know that $\log_{10} x = 4.326$ is equivalent to $10^{4.326} = x$. Solve this problem by finding the value of $10^{4.326}$. Using a calculator, we have the following.	
$x \approx 21{,}183.61135$	

Vocabulary Answers: 1. common logarithm 2. change of base formula 3. natural logarithms

Example	Student Practice
5. Evaluate antilog (-1.6784). This is the equivalent to asking what the value is of $10^{-1.6784}$. Be sure to enter the negative sign on your calculator. $10^{-1.6784} \approx 0.020970076$	**6.** Evaluate antilog (-2.5041).
7. On a scientific calculator, approximate the following values. **(a)** $\ln 7.21$ This is asking for the natural log of 7.21. Use the $\boxed{\ln}$ key on your calculator to solve. $\ln 7.21 \approx 1.975468951$ **(b)** $\ln 72.1$ $\ln 72.1 \approx 4.278054044$	**8.** On a scientific calculator, approximate the following values. **(a)** $\ln 3.25$ **(b)** $\ln 55.93$
9. On a scientific calculator, find an approximate value for x for each equation. **(a)** $\ln x = 2.9836$ If $\ln x = 2.9836$, then $e^{2.9836} = x$. Use the $\boxed{e^x}$ key on a scientific calculator to solve. Thus, $x = e^{2.9836} \approx 19.75882051$. **(b)** $\ln x = -1.5619$ If $\ln x = -1.5619$, then $e^{-1.5619} = x$. Thus, $x = e^{-1.5619} \approx 0.209737192$.	**10.** On a scientific calculator, find an approximate value for x for each equation. **(a)** $\ln x = 6.0123$ **(b)** $\ln x = -0.9774$

Example	Student Practice
11. Evaluate using common logarithms. $\log_3 5.12$	**12.** Evaluate using common logarithms. $\log_6 7.981$

Use the change of base formula. Here $b = 3$ and $x = 5.12$.

$$\log_b x = \frac{\log_a x}{\log_a b}$$

$$\log_3 5.12 = \frac{\log 5.12}{\log 3} \approx 1.486561234$$

The answer is approximate with nine decimal places, depending on the calculator you may have more or fewer digits.

13. Obtain an approximate value for $\log_4 0.005739$ using natural logarithms.	**14.** Obtain an approximate value for $\log_2 0.02546$ using natural logarithms.

Use the change of base formula where $a = e$, $b = 4$ and $x = 0.005739$.

$$\log_4 0.005739 = \frac{\log_e 0.005739}{\log_e 4}$$

$$= \frac{\ln 0.005739}{\ln 4}$$

$$\approx -3.722492455$$

To check, we want to know the following.

$$4^{-3.722492455} \overset{?}{=} 0.005739$$

Using a calculator, this can be verified using the $\boxed{y^x}$ key. The answer checks.

Example	Student Practice
15. Using a scientific calculator, graph $y = \log_2 x$.	**16.** Using a scientific calculator, graph $y = \log_4 x$.

Use the change of base formula with common logarithms to find $y = \dfrac{\log x}{\log 2}$.
Find values of y for various values of x and organize them into a data table. Graph the data points.

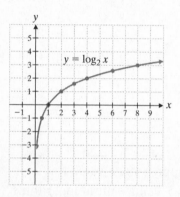

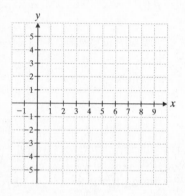

Extra Practice

1. On a scientific calculator or a graphing calculator, find a decimal approximation for $\log 11.2$.

2. On a scientific calculator, find an approximate value for x if $\ln x = 3.2$.

3. Evaluate using common logarithms. $\log_{12} 0.451$

4. Obtain an approximate value for $\log_4 0.0332$ using natural logarithms.

Concept Check
Explain how you would find x using a scientific calculator if $\ln x = 1.7821$.

Chapter 12 Logarithmic and Exponential Functions
12.5 Exponential and Logarithmic Equations

Vocabulary

logarithm   •   natural logarithm   •   logarithmic equation   •   exponential equation

1. If an exponential equation involves e raised to a power, take the _____ of each side of the equation.

2. It is not possible to take the _____ of a negative number.

3. To solve a(n) _____, get one logarithmic term on one side and a numerical value on the other, then convert it to an exponential equation using the definition of a logarithm.

Example	**Student Practice**
1. Solve. $\log_3(x+6)-\log_3(x-2)=2$	**2.** Solve. $\log_2(x+4)-\log_2(x-2)=2$

Example (continued)

Apply property 2, which states that $\log_b \dfrac{M}{N} = \log_b M - \log_b N$. Then write the equation in exponential form.

$$\log_3(x+6)-\log_3(x-2)=2$$

$$\log_3\left(\frac{x+6}{x-2}\right)=2$$

$$\frac{x+6}{x-2}=3^2$$

Solve the resulting equation.

$$\frac{x+6}{x-2}=9$$

$$x+6=9(x-2)$$

$$x+6=9x-18$$

$$24=8x$$

$$3=x$$

The check is left to the student.

Vocabulary Answers: 1. natural logarithm 2. logarithm 3. logarithmic equation

Example	Student Practice
3. Solve. $$\log(x+6)+\log(x+2)=\log(x+20)$$	**4.** Solve. $$\log(x+6)+\log(x-4)=\log(3x-4)$$

Use property 6, if $b \neq 1$, $x > 0$, $y > 0$ and $\log_b x = \log_b y$, then $x = y$.

$$\log(x+6)+\log(x+2)=\log(x+20)$$
$$\log(x+6)(x+2)=\log(x+20)$$
$$x^2+8x+12=x+20$$
$$x^2+7x-8=0$$
$$(x+8)(x-1)=0$$

$$x-1=0 \qquad x+8=0$$
$$\text{or}$$
$$x=1 \qquad\qquad x=-8$$

Check the solutions.

$$\log(1+6)+\log(1+2)\overset{?}{=}\log(1+20)$$

$$\log(7\cdot3)\overset{?}{=}\log 21$$

$$\log 21=\log 21$$

$$\log(-8+6)+\log(-8+2)\overset{?}{=}\log(-8+20)$$
$$\log(-2)+\log(-6)\neq\log(12)$$

Discard -8 because it leads to taking the logarithm of a negative number. Thus, the only solution is $x=1$.

5. Solve $2^x = 7$. Leave your answer in exact form.	**6.** Solve $5^x = 12$. Leave your answer in exact form.

Take the logarithm of each side. Solve for x.
$$2^x = 7$$

$$\log 2^x = \log 7$$
$$x \log 2 = \log 7$$
$$x = \frac{\log 7}{\log 2}$$

Example	Student Practice

7. Solve $e^{2.5x} = 8.42$. Round your answer to the nearest ten-thousandth.

$$\ln e^{2.5x} = \ln 8.42$$
$$(2.5x)(\ln e) = \ln 8.42$$
$$2.5x = \ln 8.42$$
$$x = \frac{\ln 8.42}{2.5}$$
$$x \approx 0.8522$$

8. Solve $e^{4.7x} = 10.75$. Round your answer to the nearest ten-thousandth.

---

9. If P dollars are invested in an account that earns interest at 12% compounded annually, the amount available after t years is $A = P(1+0.12)^t$. How many years will it take for $300 in this account to grow to $1500? Round your answer to the nearest whole year.

Substitute the known values and simplify.

$$1500 = 300(1+0.12)^t$$
$$1500 = 300(1.12)^t$$
$$5 = (1.12)^t$$

Now, take the common logarithm of each side and solve for t.

$$\log 5 = \log(1.12)^t$$
$$\log 5 = t(\log 1.12)$$
$$\frac{\log 5}{\log 1.12} = t$$
$$14.20150519 \approx t$$

Thus, it would take approximately 14 years.

10. If P dollars are invested in an account that earns interest at 15% compounded annually, the amount available after t years is $A = P(1+0.15)^t$. How many years will it take for $150 in this account to grow to $1700? Round your answer to the nearest whole year.

349

Example	Student Practice
11. At the beginning of 2011, the world population was seven billion people and the growth rate was 1.2% per year. If this growth rate continues, how many years will it take for the population to double to fourteen billion?	**12.** At the beginning of 2011, the mosquito population in a certain county was eight million and the growth rate was 1.7% per year. If this growth rate continues, how many years will it take for the population to double to sixteen million?

Use the formula $A = A_0 e^{rt}$ and write the population in terms of billions.

$$A = A_0 e^{rt}$$
$$14 = 7e^{(0.012)t}$$
$$2 = e^{(0.012)t}$$
$$\ln 2 = \ln e^{(0.012)t}$$
$$\ln 2 = 0.012t$$
$$57.76226505 \approx t$$

It would take about 58 years.

Extra Practice

1. Solve. $\log 2x + \log 4 = \log(2x+12)$

2. Solve. $\ln 8 - \ln x = \ln(x-7)$

3. Solve $6^x = 4^{x+2}$. Round your answer to the nearest thousandth.

4. If P dollars are invested in an account that earns interest at 6% compounded annually, the amount available after t years is $A = P(1+0.06)^t$. How many years will it take for $4000 in this account to grow to $7000? Round your answer to the nearest whole year.

Concept Check

Explain how you would solve $26 = 52e^{3x}$.

MATH COACH

Mastering the skills you need to do well on the test.

Watch the **MATH COACH** videos in MyMathLab® or on YouTube™ while you work the problems below. These helpful hints will help you avoid making common errors on test problems.

Solving an Exponential Equation—Problem 3

Solve. $4^{x+3} = 64$

> **Helpful Hint:** If one side of the equation is in exponential form, it is best to try to write the other side of the equation in exponential form. Then the procedure will be easier to complete.

Did you rewrite the equation as $4^{x+3} = 4^3$?

Yes _____ No _____

If you answered No, remember to write your equation in the form $b^x = b^y$. Note that $64 = 4^3$, and you want the base of the exponent on each side of the equation to be the same number, b, such that $b > 0$ and $b \neq 1$.

Did you use the property of exponential equations to write the equation $x + 3 = 3$?

Yes _____ No _____

If you answered No, remember that if $b^x = b^y$, then $x = y$ for any $b > 0$ and $b \neq 1$. Now solve the equation for x.

If you answered Problem 3 incorrectly, go back and rework the problem using these suggestions.

Using the Properties of Logarithms to Write Sums and Differences of Logarithms as a Single Logarithm—Problem 6   Write as a single logarithm. $2\log_7 x + \log_7 y - \log_7 4$

> **Helpful Hint:** Try your best to memorize the three properties of logarithms in Objectives 11.3.1, 11.3.2, and 11.3.3. They are essential to know when working with logarithmic expressions.

Did you use property 3 to eliminate the coefficient of the first logarithmic term, $2\log_7 x$, and rewrite that term as $\log_7 x^2$?

Yes _____ No _____

If you answered No, review property 3 in Objective 11.3.3 and complete this step again.

Did you use property 1 to combine the sum of the first two logarithmic terms and obtain the expression,

$\log_7 x^2 y - \log_7 4$?

Yes _____ No _____

If you answered No, review property 1 in Objective 11.3.1 and complete this step again.

In your last step, you will need to use property 2 in 11.3.2 to combine the difference of two logarithms.

Now go back and rework the problem using these suggestions.

Solving a Logarithmic Equation—Problem 12

Solve the equation and check your solution. $\log_8 (x+3) - \log_8 2x = \log_8 4$

> **Helpful Hint:** Use the properties of logarithms to rewrite the equation such that one logarithmic term appears on each side of the equation.

First, did you combine the two logarithms on the left side of the equation and rewrite the equation as

$$\log_8 \left(\frac{x+3}{2x} \right) = \log_8 4 ?$$

Yes _____ No _____

If you answered No, use property 2 from Objective 11.3.2 to complete this first step again.

Next, did you rewrite the equation as $\dfrac{x+3}{2x} = 4 ?$

Yes _____ No _____

If you answered No, notice that you now have one logarithmic term on each side of the equation, which is the goal mentioned in the Helpful Hint. You can use property 6 from Objective 11.3.4 to evaluate these two logarithms.

In your final step, solve the equation for x. Check your solution by substituting for x in the original equation, evaluating the logarithms and then simplifying the resulting equation.

If you answered Problem 12 incorrectly, go back and rework the problem using these suggestions.

Solving an Exponential Equation Involving e Raised to a Power—Problem 14

Solve the equation. Leave your answer in exact form. Do not approximate. $e^{5x-3} = 57$

> **Helpful Hint:** The first step is to take the natural logarithm of each side of the equation. Then you can simplify the equation further using the properties of logarithms.

Did you take the natural logarithm of each side of the equation to obtain $\ln e^{5x-3} = \ln 57 ?$
Yes _____ No _____

Next did you rewrite this equation as $(5x-3)(\ln e) = \ln 57 ?$
Yes _____ No _____

If you answered No to these questions, review property 3 of logarithms from Objective 11.3.3 and complete this step again.

Did you rewrite the equation as $5x - 3 = \ln 57 ?$
Yes _____ No _____

If you answered No, remember that $\ln e = \log_e e = 1$. This combines the definition of natural logarithms or $\ln e = \log_e e$ and property 4 of logarithms from Objective 11.3.4, which indicates that $\log_e e = 1$.

In your final step, solve the equation for x without evaluating $\ln 57$.

Now go back and rework the problem using these suggestions.

0.1

Student Practice

2.   (a) $\dfrac{11}{13}$

    (b) $\dfrac{3}{5}$

4. $\dfrac{1}{13}$

6. 9

8. $\dfrac{3}{5}$

10. 9

12. $\dfrac{77}{12}$

14. (a) 11
    (b) 20

Extra Practice

1. 4

2. $55\dfrac{3}{5}$

3. $\dfrac{29}{17}$

4. 64

Concept Check

Answers may vary. Possible solution:
First, multiply the whole number by
the denominator. Then, add this to the
numerator. The result is the new
numerator. The denominator does not
change.

0.2

Student Practice

2. (a) 1

    (b) $\dfrac{3}{2}$ or $1\dfrac{1}{2}$

4. $\dfrac{1}{4}$

6. 70

8. 126

10. $\dfrac{95}{108}$

12. $47\dfrac{1}{3}$ yd

Extra Practice

1. 100

2. $\dfrac{5}{9}$

3. $3\dfrac{7}{36}$

4. $3\dfrac{1}{2}$

Concept Check

Answers may vary. Possible solution:
Write each denominator as the product
of prime factors. The LCD is a product
containing each different factor. If a
factor occurs more than once in any one
denominator, the LCD will contain that
factor repeated the greatest number of
times that it occurs in any one
denominator.

0.3

Student Practice

2. (a) $\dfrac{2}{17}$

    (b) $\dfrac{11}{27}$

4. 20

6. $44\dfrac{5}{8}$ m^2

8. $\dfrac{2}{15}$

10. 35

12. $\dfrac{3}{8}$

14. $3\dfrac{4}{7}$ or $\dfrac{25}{7}$

16. $\frac{4}{5}$ image per hour

Extra Practice

1. $\frac{2}{49}$

2. 18

3. $\frac{3}{2}$ or $1\frac{1}{2}$

4. 14 pieces

Concept Check

Answers may vary. Possible solution: Change each mixed number to an improper fraction. Invert the second fraction and multiply the result by the first fraction. Simplify.

0.4

Student Practice

2. (a) $\frac{9}{100,000}$; 5 decimal places; nine hundred-thousandths

   (b) $4\frac{25}{1000}$; 3 decimal places, four and twenty-five thousandths

4. $0.8666...$ or $0.8\overline{6}$

6. $\frac{18}{25}$

8. 37.11

10. 0.000425

12. 84,000

14. 93

16. 713.83

Extra Practice

1. $32\frac{41}{500}$; thirty-two and eighty-two thousandths

2. 6.1368

3. 19.32

4. 1678.5

Concept Check

Answers may vary. Possible solution: Move the decimal point over 4 places to the right on the divisor $(0.0035 \Rightarrow 35)$.

Then move the decimal point the same number of places on the dividend $(0.252 \Rightarrow 2520)$.

0.5

Student Practice

2. (a) 0.019%
   (b) 60%

4. (a) 595%
   (b) 790%

6. (a) 0.008
   (b) 1.012

8. 31.95

10. (a) $79.95
    (b) $43.05

12. 87.5%

14. 6%

16. 1000 square meters

Extra Practice

1. 0.576%

2. 1

3. 260%

4. 0.02

Concept Check

Answers may vary. Possible solution: Move the decimal point two places to the right and add the % symbol.

0.6

Student Practice

2. $650

4. (a) 18%
   (b) 28%
   (c) The Northwest

Extra Practice

1. 492 people

2. $701.25

3. 1.7 kilometers

4. $3380

Concept Check

Answers may vary. Possible solution: Multiply the number of kilometers by 0.62 to obtain miles.

1.1

Student Practice

2.   (a) Irrational, real
     (b) Rational, real
     (c) Rational, real

4.   (a) −6.23
     (b) +109.4
     (c) −345
     (d) +15

6.   (a) $\dfrac{7}{8}$
     (b) 30 feet above sea level

8.   (a) 5.34
     (b) $\sqrt{5}$
     (c) 0

10. −14

12. −6

14. 0.1

16. −8

Extra Practice

1.   Irrational, real

2. $\dfrac{7}{8}$

3.   −25

4.   15

Concept Check

Answers may vary. Possible solution: Adding a negative number to a negative number will always result in a number further in the negative direction. However, adding numbers of opposite sign could result in a negative number if the absolute value of the negative number is larger than that of the positive number. Further, adding numbers of opposite sign could result in a positive number, if the absolute value of the positive number is greater than that of the negative number. Finally, adding numbers of opposite sign could result in 0 if their absolute values are equal.

1.2

Student Practice

2.   12

4.   −6

6.   (a) $-\dfrac{5}{9}$
     (b) $-\dfrac{3}{20}$

8.   0

10. (a) −20
     (b) −5
     (c) 25
     (d) $-\dfrac{26}{5}$

12. 629 feet

Extra Practice

1.   −3.29

2. $-\dfrac{7}{30}$

3.   −11

4. $179-(-22)$; 201 feet

Concept Check

Answers may vary. Possible solution: The result could be positive if the absolute value of the number subtracted is greater than the other number

$$\left[-2-(-3)=-2+3=1\right].$$

The result could be zero if the two numbers are the same

$$\left[-2-(-2)=-2+2=0\right].$$

The result could be negative if the absolute value of the number being subtracted is less than the other number

$$\left[-2-(-1)=-2+1=-1\right].$$

1.3

Student Practice

2.   (a) 20
     (b) $-\dfrac{25}{12}$ or $-2\dfrac{1}{12}$

4. (a) 32.4
   (b) −60
6. (a) 8
   (b) −5
8. (a) 11
   (b) −0.6
10. $\dfrac{5}{12}$
12. −40

Extra Practice
1. −462.94
2. 16
3. −175
4. $-\dfrac{5}{6}$

Concept Check
Answers may vary. Possible solution:
Multiplying an even number of negative numbers results in a positive number, whereas multiplying an odd number of negative numbers results in a negative number.

1.4
Student Practice
2. (a) $(-3)^4$
   (b) $(-6)^8$
   (c) $(n)^6$
4. (a) 64
   (b) 37
6. (a) −64
   (b) 256
   (c) −256
   (d) −256
8. (a) $\dfrac{1}{125}$
   (b) 0.027
   (c) $\dfrac{64}{343}$
   (d) 400
   (e) −24

Extra Practice
1. $(-ab)^2$

2. $-\dfrac{1}{8}$
3. −40
4. 57

Concept Check
Answers may vary. Possible solution:
If you have parentheses surrounding the −2, then the base is −2 and the exponent is 6. The result is 64. If you do not have parentheses, then the base is 2. You evaluate to obtain 64 and then take the opposite of 64, which is −64. Thus, $(-2)^6 = 64$ and $-2^6 = -64$.

1.5
Student Practice
2. 60
4. 24
6. 23
8. $-\dfrac{5}{48}$

Extra Practice
1. 1
2. 29
3. 2
4. $-\dfrac{31}{5}$

Concept Check
Answers may vary. Possible solution:
Evaluate the operations within the parentheses. Then raise the result to a power of 3. Then divide. Finally, add and subtract from left to right.

1.6
Student Practice
2. (a) $6x + 18y$
   (b) $-8m - 20n$
4. $-m + 6n$
6. (a) $\dfrac{3}{4}x^2 - 6x + \dfrac{15}{4}$
   (b) $1.3x^2 + 1.69x + 0.91$
8. $-5x^2 - 15xy + 30x$

10. $-8y+6y^2$

12. $3500x+4000y$

Extra Practice

1. $-12x-6$

2. $\dfrac{1}{6}x+\dfrac{1}{5}y$

3. $-18x+12y-6z$

4. $30(15+7n);\ 450+210n$

Concept Check

Answers may vary. Possible solution:

Distribute $\left(-\dfrac{3}{7}\right)$ to each term within

the parentheses.

$$\left(-\dfrac{3}{7}\right)\left(21x^2-14x+3\right)$$

$$=\left(-\dfrac{3}{7}\right)\left(21x^2\right)+\left(-\dfrac{3}{7}\right)(-14x)+\left(-\dfrac{3}{7}\right)(3)$$

$$=-9x^2+6x-\dfrac{9}{7}$$

1.7

Student Practice

2. (a) $4x$ and $2x$

   (b) $4x^2$ and $-6x^2$; $5x$ and $9x$

4. (a) $7z^5$

   (b) $2z$

6. $3x^3$

8. (a) $7.6m+7.2n$

   (b) $8a^2b-4ab^3$

   (c) $6xy+2x^2y-9xy^2+5x^2y^2$

10. $14x-5c-2t$

12. $\dfrac{17}{5}x-\dfrac{5}{14}y$

14. $-17c+47cd$

Extra Practice

1. $-2a+12+4a^2$

2. $\dfrac{9}{14}x^2-\dfrac{5}{8}y$

3. $-22a+22b$

4. $2ab+13ac-3bc+30cd$

Concept Check

Answers may vary. Possible solution:
Use the distributive property to remove
the parentheses. Then simplify by
combining like terms.

$$1.2(3.5x-2.2y)-4.5(2.0x+1.5y)$$

$$=4x-2.64y-9x-6.75y$$

$$=-4.8x-9.39y$$

1.8

Student Practice

2. -25

4. (a) 75

   (b) 225

6. 33

8. 118

10. 13.5

12. 50.24 yd^2

14. $-40°$ C

16. You are driving at approximately
    112.7 km/h. You are exceeding
    the speed limit.

Extra Practice

1. -3

2. 19

3. -1

4. 113.04 m^2

Concept Check

Answers may vary. Possible solution:
The formula for the area of a circle is
$A=\pi r^2$. Therefore, first find the radius

from the diameter $\left(r=\dfrac{d}{2}\right)$.

Use $\pi\approx3.14$.

$$A=3.14(6\text{ m})^2$$

$$=3.14\left(36\text{ m}^2\right)=113.04\text{ m}^2$$

1.9

Student Practice

2. $4x-60$

4. $-32x+8y+8$

6. $32x-42$

8. $-8x-8$

357

Extra Practice

1. $-5a-9b$
2. $10y-2x$
3. $a-6b$
4. $-10x^2-16x+20$

Concept Check

Answers may vary. Possible solution: First combine like terms within the square brackets. Then use the distributive property to remove grouping symbols and combine like terms after each step.

$$3\{2-3[4x-2(x+3)+5x]\}$$
$$=3\{2-3[9x-2(x+3)]\}$$
$$=3\{2-3[9x-2x-6]\}$$
$$=3\{2-3[7x-6]\}$$
$$=3\{2-21x+18\}$$
$$=3\{20-21x\}$$
$$=60-63x$$

Worksheet Answers Chapter 2

2.1

Student Practice

2. $x = 34$
4. $x = 12$
6. $x = 2.3$
8. No. $x = 36$
10. $x = -\dfrac{5}{21}$

Extra Practice

1. $x = 6$
2. $x = 3.53$
3. No. $x = -6$
4. Yes.

Concept Check

Answers may vary. Possible solution: Substitute the value 3.8 for x in the equation. Simplify. If the resultant equation is true, $x = 3.8$ is the solution. If the resultant equation is not true, $x = 3.8$ is not the solution.

2.2

Student Practice

2. $x = 21$
4. $x = 7$
6. $x = \dfrac{15}{8}$
8. $x = -7$
10. $x = -15$
12. $x = -6$
14. $x = 7$

Extra Practice

1. $x = 13$
2. $x = -25$
3. Yes
4. No; $x = -0.12$

Concept Check

Answers may vary. Possible solution:

Change $36\dfrac{2}{3}$ to an improper fraction.

Substitute that value for x in the equation. Simplify. If the resultant

equation is true, $x = 36\dfrac{2}{3}$ is the

solution.

2.3

Student Practice

2. $x = 2$
4. $x = -\dfrac{2}{3}$
6. $x = \dfrac{17}{4}$
8. $x = 1$
10 $z = 20$

Extra Practice

1. $x = 5$
2. $x = 12$
3. $x = \dfrac{1}{4}$
4. $z = 2$

Concept Check

Answers may vary. Possible solution: Use the distributive property to remove parentheses. Combine like terms on the left side of the equation. Move the variable terms to the left side of the equation and the constants to the right side of the equation. Simplify.

2.4

Student Practice

2. $x = 3$
4. $x = -15$
6. $x = 112$
8. $x = \dfrac{1}{3}$

Extra Practice

1. $p = \dfrac{3}{2}$
2. $x = -2$
3. $x = -5$
4. $x = -5$

359
Copyright © 2013 Pearson Education, Inc.

Concept Check

Answers may vary. Possible solution: Multiply both sides of the equation by the LCD, 12. Add or subtract terms on both sides of the equation to get all terms containing x on one side of the equation. Add or subtract a constant value to both sides of the equation to get all terms not containing x on the other side of the equation. Divide both sides by the coefficient of x and simplify the solution if necessary. Finally, check the solution.

2.5

Student Practice

2. (a) $x - 5$

    (b) $3x$

    (c) $\dfrac{x}{5}$ or $\dfrac{1}{5}x$

    (d) $x + 10$

4. a. $3x + 8$

    b. $3(x + 8)$

    c. $\dfrac{1}{4}(x + 6)$

6. $c =$ Chuck's car's mileage; $c + 12{,}000 =$ Tom's car's mileage

8. $w =$ the width; $3w - 7 =$ the length

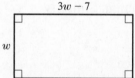

10. $s =$ the second angle; $5s =$ the first angle; $s + 40 =$ the third angle

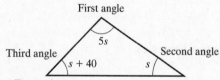

Extra Practice

1. $x + 12$

2. $\dfrac{1}{4}x - 1$

3. $x =$ value of Allison's car; $x + \$2300 =$ value of Alicia's car

4. $s =$ number of Scott's comic books; $s + 17 =$ number of Patricia's comic books; $3s =$ number of Walter's comic books; $4s - 5 =$ number of Adrienne's comic books

Concept Check

Answers may vary. Possible solution: "one-third of the sum" means you multiply $\dfrac{1}{3}$ times the sum. The sum will be the quantity in the parentheses because the $\dfrac{1}{3}$ is multiplied by the whole sum; $\dfrac{1}{3}(x + 7)$.

2.6

Student Practice

2. 85

4. 65

6. $32°$

8. 12 hours for John; 13 hours for Yuri

Extra Practice

1. 546

2. 15

3. 18 energy bars

4. 300 minutes

Concept Check

Answers may vary. Possible solution: Since we want to know how many pairs of socks, let $x =$ the number of pairs of socks. Set up an equation to represent the total amount spent.

$2(\$23) + \$0.75x = \$60.25$

2.7

Student Practice

2. 16 hours

4. $800
6. $2000 invested at 6%; $1500 invested at 5%
8.   21 nickels; 17 dimes

Extra Practice
1. $16,800
2. $975
3. $54,000
4.   129 bills

Concept Check
Answers may vary. Possible solution: Since each amount is given in terms of quarters, let $x=$ the number of quarters. Then $2x=$ the number of dimes, and $x+1=$ the number of nickels. Now set up an equation and solve, given the value of each coin and the total.
$$0.25x+0.10(2x)+0.05(x+1)=2.55$$

Extra Practice
1. $<$
2.
3. $w \geq 1,000,000$
4. $x < 3$
   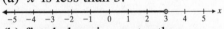

Concept Check
Answers may vary. Possible solution: $12 < x$ is the same as $x > 12$, but written in different ways. The graphs will be exactly the same.

2.8

Student Practice
      (a) $>$
      (b) $>$
      (c) $<$
      (d) $>$
2.   (e) $<$
4.   (a) x is less than 3.
   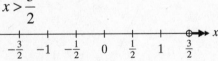
      (b) five halves is greater than or equal to x.
   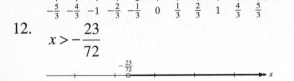
6.   (a) $c < \$125$
      (b) $p \leq 35$

8. $x > \dfrac{3}{2}$

10. $x > -\dfrac{1}{3}$

12. $x > -\dfrac{23}{72}$

Worksheet Answers Chapter 3

3.1

Student Practice

2.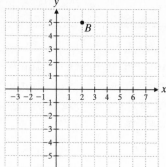

4.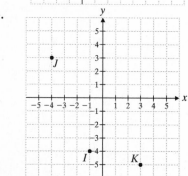

6. $(-3,6)$

8. No

10. $(-2,3)$

Extra Practice

1.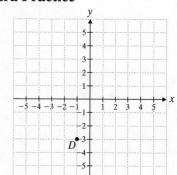

2. $(3,5)$

3. (a) $(-1,5)$

   (b) $(3,-3)$

4. (a) $(-1,-8)$

   (b) $(4,2)$

Concept Check

Answers may vary. Possible solution: Isolate x on the left side of the equation. Substitute the given value for y and solve for x.

3.2

Student Practice

2.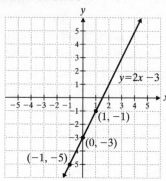

4.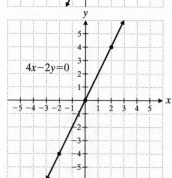

6. (a) x-intercept:$(-1,0)$

   y-intercept:$(0,-2)$

   (b)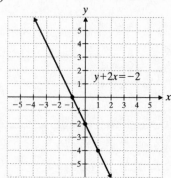

8.

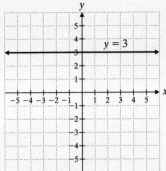

$y = 3$

Extra Practice

1.

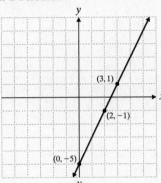

$(3, 1)$
$(2, -1)$
$(0, -5)$

2.

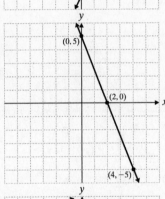

$(0, 5)$
$(2, 0)$
$(4, -5)$

3.

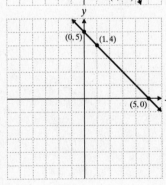

$(0, 5)$ $(1, 4)$
$(5, 0)$

4.

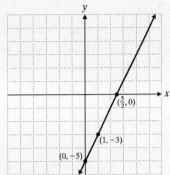

$\left(\frac{5}{2}, 0\right)$
$(1, -3)$
$(0, -5)$

Concept Check

Answers may vary. Possible solution: The most important ordered pair is $(0, 0)$, since it is both the x- and y- intercept of the equation.

3.3

Student Practice

2. $m = \dfrac{2}{3}$

4. (a) No slope
    (b) $m = 0$

6. $m = -3$, y-intercept: $(0, 4)$

8.
    (a) $y = \dfrac{3}{4}x - 7$
    (b) $3x - 4y = 28$

10.

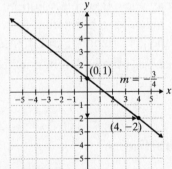

$(0, 1)$
$m = -\dfrac{3}{4}$
$(4, -2)$

Extra Practice

1. $-\dfrac{1}{3}$

2. $m = 5$, y-intercept: $(0, 0)$

3. $y = -4x + \dfrac{4}{5}$

4. $y = -2$

Concept Check
Answers may vary. Possible solution:
Slope measures the vertical change per
one unit of horizontal change.

3.4
Student Practice

2. $y = -\dfrac{1}{2}x - 4$

4. $y = -\dfrac{3}{5}x + 7$

6. $y = -\dfrac{5}{2}x + 5$

8.
   (a) $\dfrac{5}{7}$

   (b) $-\dfrac{7}{5}$

Extra Practice

1. $y = -\dfrac{1}{2}x + 2.5$

2. $y = 5x - 11$

3. $y = \dfrac{7}{6}x - 1$

4. $y = -5x + 8$

Concept Check
Answers may vary. Possible solution:
Zero slope indicates a horizontal line.
Since the line is horizontal, and it passes
through $(-2, -3)$, the equation must be
$y = -3$.

3.5
Student Practice

2.
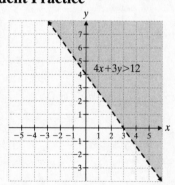

4.
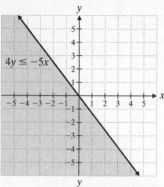

6.
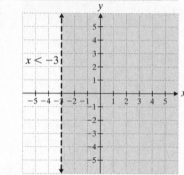

Extra Practice

1.
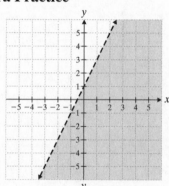

2.
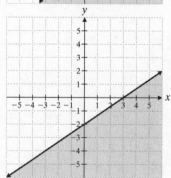

3.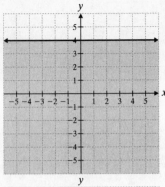

4.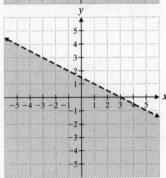

Concept Check
Answers may vary. Possible solution: The inequality is first graphed without shading. The test point coordinates are then substituted into the inequality. If the result of the substitution results in a true statement, the area where the test point lies is shaded. If false, the opposite area is shaded.

8. (a) Not a function
    (b) Function
10. (a) 49
    (b) 19
    (c) 7

Extra Practice
1. The domain is $\{2.5, 3.5, 5.5, 8.5\}$.

    The range is $\{-6, -2, 0, 3\}$.

    Function
2. (a) 1
    (b) 28
    (c) 10
3. Not a function
4.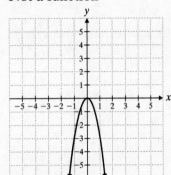

Concept Check
Answers may vary. Possible solution: Duplicate elements are recorded only once in the domain and in the range.

3.6
Student Practice
2. The domain is $\{-3, -2, 4, 7\}$.

    The range is $\{1, 6\}$.

4. (a) Not a Function
    (b) Function
6.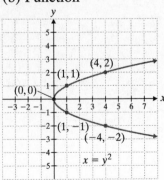

4.1

Student Practice

2. $(1,1)$

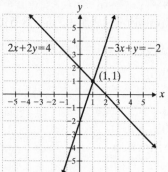

4. No solution; inconsistent system

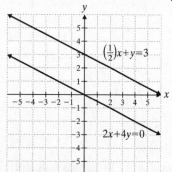

6. Infinite number of solutions; dependent system

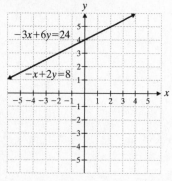

8.
(a)
$$y = 50 + 40x$$
$$y = 90 + 30x$$

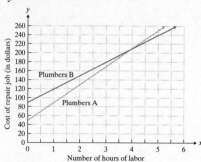

(b)

(c) 4 hours

(d) Plumbers B

Extra Practice

1. $(-2, -6)$

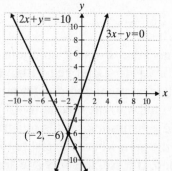

2. No solution; inconsistent system

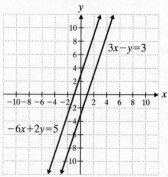

366

3. $(6,2)$

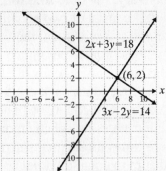

4. Infinite number of solutions; dependent system

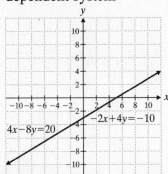

Concept Check
Answers may vary. Possible solution: The two equations represent lines that are parallel lines. There is, therefore, no solution.

4.2

Student Practice
2. $(-2,3)$
4. $\left(\dfrac{3}{8},\dfrac{5}{8}\right)$

Extra Practice
1. $(0,-3)$
2. $\left(\dfrac{54}{5},\dfrac{6}{5}\right)$
3. $(-6,6)$
4. $(-45,33)$

Concept Check
Answers may vary. Possible solution: Label the equations (1) and (2).

Solve (1) for y and label this equation (3). Substitute the value of y in (3) into (2) and solve (2) for x. Substitute the found value of x back into (3) and solve for y. Check that the values of x and y satisfy both (1) and (2).

4.3

Student Practice
2. $(4,-1)$
4. $(2,3)$
6. $(-3,1)$

Extra Practice
1. $(29,23)$
2. $(-7,-4)$
3. $\left(1,\dfrac{4}{3}\right)$
4. $\left(\dfrac{1}{2},-2\right)$

Concept Check
Answers may vary. Possible solution: Multiplying both equations by 100 will eliminate decimals. Or, multiply the first equation by 10 and the second equation by 100.

4.4

Student Practice
2. $(15,8)$
4. No solution; inconsistent system
6. Infinite number of solutions; dependent system

Extra Practice
1. $(-4,-5)$
2. No solution; inconsistent system

3. $(3,5)$

4. $(-3,-5)$

Concept Check

Answers may vary. Possible solution:
Label the equations (1) and (2).

Multiply (1) by 10 and (2) by -48.

Then add equations to eliminate
fractions and x.

4.5

Student Practice

2. The first employee plants 15 bulbs
   per hour and the second employee
   plants 20 bulbs per hour.

4. The student should use 50
   milliliters of the 25% solution and
   100 milliliters of the 10% solution.

6. The plane's speed in still air was 80
   nautical miles per hour and the
   wind speed was 20 nautical miles
   per hour.

Extra Practice

1. They have 17 quarters and 9
   nickels.

2. They should include 5 pounds of
   large snack bars and 5 pounds of
   small snack bars.

3. Ronaldo hammers 85 nails per hour
   and Lee hammers 75 nails per hour.

4. It will be 25 years until the
   populations are the same. Each
   population will be 55,600 people.

Concept Check

Answers may vary. Possible solution:
The system will consist of two equations
in terms of x and y. x will equal the
number of gallons of 50% pure juice. y
will equal the number of gallons of 30%
pure juice. The first equation states that
the sum of x and y will equal 500. The
second equation states that each amount
of the mixture times its decimal fraction
of juice will sum to equal the total
mixture quantity times its decimal
fraction of juice.

5.1

Student Practice

2. (a) z^5

    (b) 2^8

4. $-8x^6y^3$

6. x^4

8. $\dfrac{1}{n^4}$

10. $\dfrac{x^2}{2y^5}$

12. (a) y^{32}

    (b) -1

14. $\dfrac{y^3}{z^6}$

16. $-\dfrac{x^{15}}{64y^6}$

Extra Practice

1. $24x^6y^5$

2. 9

3. $\dfrac{1}{b^4}$

4. $\dfrac{9^2}{14^2x^3}$, or $\dfrac{81}{196x^6}$

Concept Check

Answers may vary. Possible solution:
In the numerator and the denominator,
raise each factor inside the
parentheses to the power.

$$\dfrac{4^2\left(x^3\right)^2}{2^3\left(x^4\right)^3}$$

Evaluate the constants, and multiply
exponents on the variable expressions.

$$\dfrac{16x^6}{8x^{12}}$$

Divide the numbers and subtract
exponents on the variable expressions.
Because the larger exponent is in the
denominator, the variable expression

will be in the denominator, $\dfrac{2}{x^6}$.

5.2

Student Practice

2. $\dfrac{1}{2401}$

4. $\dfrac{y^8}{16x^4z^{12}}$

6. 2.564×10^3

8. (a) 1.3×10^{-3}

    (b) 1.0×10^{-6}

10. (a) 0.000528

    (b) $3,221,400$

12. 3.86×10^{11} hr

Extra Practice

1. $\dfrac{1}{343}$

2. $\dfrac{b^8c^{12}}{81a^{12}}$

3. 6340

4. 2.0×10^{-1}

Concept Check

Answers may vary. Possible solution:
Raise each factor inside the parentheses
to the power.

$$4^{-3}\left(x^{-3}\right)^{-3}\left(y^4\right)^{-3}$$

Multiply exponents on the variable
expressions.

$$4^{-3}x^9y^{-12}$$

Rewrite as a fraction.

$$\dfrac{x^9}{4^3y^{12}}$$

Evaluate 4^3.

$$\dfrac{x^9}{64y^{12}}$$

5.3

Student Practice

2. (a) Degree 7; monomial
   (b) Degree 2; trinomial
   (c) Degree 7; binomial

4. $4x^2 - 8x + 5$

6. $\dfrac{7}{4}x^2 - \dfrac{8}{5}x + \dfrac{5}{8}$

8. $-3x^2 - 10x + 5$

10. 26.4 miles per gallon

Extra Practice

1. Degree 6; trinomial

2. $10.8x - 14$

3. $5r^4 - r^2 + 8$

4. 9.01 miles per gallon

Concept Check

Answers may vary. Possible solution:
To determine the degree of the
polynomial, first determine the degree
of each term by finding the sum of the
exponents on the variables in each
term. The degree of the first term,
$2xy^2$, is $1 + 2 = 3$, and the degree of the
second term, $-5x^3y^4$, is $3 + 4 = 7$. The
degree of the polynomial is the greater
of these, which is 7. To determine
whether the polynomial is a monomial,
a binomial, or a trinomial, we must
count the number of terms in the
polynomial. The polynomial has two
terms, $2xy^2$ and $-5x^3y^4$, so it is a
binomial.

5.4

Student Practice

2. $-5y^3 + 20y^2$

4. $2x^4y - 14x^3y - 20x^2y$

6. $15x^2 + 4x - 3$

8. $7x^2 + 14xz - 3xy - 6yz$

10. $16x^2 + 24xy + 9y^2$

12. $20x^4 - 20x^2y^4 - 15y^8$

14. $8x^2 - 2$

Extra Practice

1. $-10x^4 + 3x^2$

2. $x^2 - 14x + 33$

3. $15x^2 - 11xy - 56y^2$

4. $8x^4 - 2x^2y^3 - 15y^6$

Concept Check

Answers may vary. Possible solution:
First write the square of the binomial as
the product of the binomial and itself.
$(7x - 3)(7x - 3)$
Then use FOIL and collect like terms.
$49x^2 - 21x - 21x + 9 = 49x^2 - 42x + 9$

5.5

Student Practice

2. $64y^2 - 16$

4. $4x^2 - 81y^2$

6. $49x^2 + 14xy + y^2$

8. $2x^4 + 2x^3 - 29x^2 + 9x + 30$

10. $x^4 - 7x^3 + 6x^2 - 22x + 40$

12. $2x^3 + 6x^2 - 8x - 24$

Extra Practice

1. $x^2 - 100$

2. $25a^2 - 9b^2$

3. $24x^3 + 2x^2 + 7x + 3$

4. $x^3 - 4x^2 - 7x + 10$

Concept Check

Answers may vary. Possible solution:
To use the formula
$(a + b)^2 = a^2 + 2ab + b^2$ to multiply
$(6x - 9y)^2$, first identify a and b: $a = 6x$
and $b = -9y$. Then substitute these
values for a and b in the formula and
simplify.

Math displayed on next page.

370

$$(6x-9y)^2$$

$$=(6x)^2 + 2(6x)(-9y) + (-9y)^2$$

$$=36x^2 - 108xy + 81y^2$$

5.6

Student Practice

2. $3x^3 + x - 4$

4.
$$2x^2 + 2x + 1 + \frac{5}{x+1}$$

6.
$$4x^2 - 2x - 3 + \frac{8}{2x+1}$$

8.
$$4x^2 + 4x + 6 + \frac{24}{2x-3}$$

Extra Practice

1. $6a^5 - 2a^3 + 4a - 1$

2. $4y^2 - 6y + 9$

3. $3x^2 - 2x + 5$

4.
$$y^2 + y - 1 - \frac{5}{y-1}$$

Concept Check

Answers may vary. Possible solution:
Multiply the quotient and the divisor.
Then add the remainder. You should
get the original dividend.

$$(x-2)\left(x^2 + 2x + 8\right) + 13$$

$$= x^3 + 2x^2 + 8x - 2x^2 - 4x - 16 + 13$$

$$= x^3 + 4x - 3$$

Yes, the answer checks.

6.1

Student Practice

2.   (a) $5(y+3z)$

    (b) $y(12-5z)$

4. $11y(4y^2+5y-x)$

6.   (a) $7(3m^2-4n^2)$

    (b) $m^2(mn^2+9n^2+3m^2)$

8. $9xy^2(3y-4x-x^2y)$

10. $11m^3n^2(m+1)$

12. $(3x+4y)(5y^2-1)$

14. $\pi(m^2+4n^2+9z^2)$

Extra Practice

1. $3x(x^2+4x-7)$

2. $6x(10xy+3y-4)$

3. $(x+3y)(8a-b)$

4. $(5a-1)(4x-3)$

Concept Check
Answers may vary. Possible solution:
Determine that the largest integer that
will divide into the coefficient of all
terms is 36. Determine that the
variables common to all terms are a^2
and b^2.
Write the above common factors as
the first part of the answer (the first
factor).
Remove common factors, and what
remains is the second part of the
answer (the second factor).

6.2

Student Practice

2. $(2z+5)(z-4)$

4. $(3x+1)(4x+5)$

6. $(y+3z)(7+x)$

8. $(z+y)(n+6)$

10. $(2x+3)(x-3)$

12. $(5z-1)(n-m)$

14. $(5x-3w)(7y-3z)$

Extra Practice

1. $(x-2)(x+1)$

2. $(x+5)(x-2)$

3. $(3x-2)(2x+5)$

4. $(3a+2b)(4x+5y)$

Concept Check
Answers may vary. Possible solution:
Start by grouping terms $10ax$ with $5ab$
and $2bx$ with b^2.
$(2x+b)$ can be factored out of both
groups leaving the second factor to be
$(5a+b)$.

6.3

Student Practice

2. $(x+3)(x+6)$

4. $(x+11)(x+3)$

6. $(x-4)(x-7)$

8. $(x-6)(x+4)$

10. $(z-2)(z+14)$

12. $(x^2+2)(x^2-5)$

14. $4(x-7)(x+3)$

16. $(x+10)(x-7)$

Extra Practice

1. $(x+2)(x+4)$

2. $(a-3)(a-4)$

3. $(x-4)(x+3)$

4. $2(x+2)(x+7)$

Concept Check
Answers may vary. Possible solution:
The first step is to factor out the
greatest common factor of 4, leaving
$4\left(x^2 - x - 30\right)$. Next write the
expression in factored form using
variables m and n, $4(x+m)(x+n)$.
Next determine that the product of m
and n is -30, and the sum is -1.
m and n may equal 5 and -6.
Substitute the values of m and n, then
check.

6.4
Student Practice

2. $(2x+5)(x+4)$

4. $(3x-1)(x-1)$

6. $(x-3)(4x+7)$

8. $(x+1)(2x+3)$

10. $(x-3)(5x+2)$

12. $2(4x-1)(x+3)$

14. $5(2x-3)(3x-2)$

Extra Practice

1. $(2x-1)(2x-5)$

2. $(7x+3)(x-1)$

3. $(5x-2)(x+9)$

4. $3(5y-3)(y+4)$

Concept Check
Answers may vary. Possible solution:
First step is to factor out common
coefficients and variables.
$2x\left(5x^2 + 9xy - 2y^2\right)$

Next step is to factor the inside
expression using grouping.
The grouping number is -10.
$2x\left[5x^2 + 10xy - xy - 2y^2\right]$

$= 2x\left[5x(x+2y) - y(x+2y)\right]$

$= 2x(2x - y)(x+2y)$

Lastly, check the solution by multiplying
the factors.

6.5
Student Practice

2. $(6x+1)(6x-1)$

4. $(8x+3)(8x-3)$

6. $\left(4x^2+1\right)(2x+1)(2x-1)$

8. $(x+5)^2$

10. (a) $(8x+5y)^2$

(b) $\left(7x^2-1\right)^2$

12. $(4x+1)(9x+4)$

14. $3(x+5)(x-5)$

16. $3(4x-3)^2$

Extra Practice

1. $(5a-6b)(5a+6b)$

2. $(6a+5b)^2$

3. $\left(x^2-10\right)\left(x^2+10\right)$

4. $\left(3x^2-5\right)^2$

Concept Check
Answers may vary. Possible solution:
First factor out the common factor of 2.
$2\left(12x^2 + 60x + 75\right)$

Next use grouping to factor the inside
expression. The grouping number is 900.
$2\left(12x^2 + 30x + 30x + 75\right)$

$= 2\left[6x(2x+5) + 15(2x+5)\right]$

$= 2(6x+15)(2x+5)$

Lastly, check by multiplying.

6.6
Student Practice

2. (a) $y(3y+1)^2$

(b) $-4x(x-8)(x+1)$

4. $(b-3)(x-4)(x+4)$

6. (a) prime
    (b) prime

Extra Practice

1. prime

2. $7x(3-x)(3+x)$

3. $-x(x+3)(x-15)$

4. $(x+3)(x+2)$

Concept Check

Answers may vary. Possible solution:
The first step is to group terms and factor out common factors.

$2x(x+3w)-5(x+3w)$

Next, factor out $(x+3w)$.

$(x+3w)(2x-5)$

Finally, check by multiplying.

6.7

Student Practice

2. $\dfrac{1}{3}$ and -3

4. 0 and $\dfrac{4}{5}$

6. $x=7,\ x=-9$

8. width $=12$ in., length $=5$ in.

10. 3 seconds

Extra Practice

1. $3,\ -\dfrac{1}{2}$

2. $-3,3$

3. $7,\ -5$

4. $3,4$

Concept Check

Answers may vary. Possible solution:
Let $x=$ width of rectangle, then
length $=(2x+3)$.

$A=(\text{width})(\text{length})$

$65=x(2x+3)$

$65=2x^2+3x$

$0=2x^2+3x-65$

$0=(2x+13)(x-5)$

Set each factor equal to 0 and solve for x.

$$2x+13=0 \qquad x-5=0$$
$$2x=-13 \qquad x=5$$
$$x=-\frac{13}{2}$$

x cannot be negative in this case, because it describes a length, So $x=5$.

length $=2(5)+3=13$ feet

width $=5$ feet

Worksheet Answers Chapter 7

7.1

Student Practice

2. $\dfrac{8}{11}$

4. $\dfrac{2}{5}$

6. $\dfrac{x+7}{x+4}$

8. $\dfrac{x+2}{x-6}$

10. $-\dfrac{4}{7}$

12. $-\dfrac{3x+4}{5+2x}$

14. $\dfrac{5x+3y}{9x+7y}$

16. $\dfrac{5a+4b}{2a+b}$

Extra Practice

1. $\dfrac{3}{x}$

2. $\dfrac{3x-4}{3x+4}$

3. $-\dfrac{x+9}{3(x+2)}$

4. $\dfrac{5x-2y}{3x+y}$

Concept Check

Answers may vary. Possible solution: Completely factoring both numerator and denominator is the only way to see what factors are shared, and may consequently be eliminated. In this case, it can be seen that $(x-y)$ is a common factor.

7.2

Student Practice

2. $\dfrac{3x+7}{2x-3}$

4. $\dfrac{x}{(x+5)(3x+4)}$ or $\dfrac{x}{3x^2+19x+20}$

6. $\dfrac{4x-3}{2x+1}$

8. $-\dfrac{1}{(x-5)(x-2)}$

Extra Practice

1. $\dfrac{x}{x+1}$

2. $\dfrac{x}{5x+7}$

3. $\dfrac{1}{(x-2)(x+5)}$

4. $\dfrac{5(x-1)}{3x+4}$

Concept Check

Answers may vary. Possible solution: The first step is to change the operation from division to multiplication, by changing the operator and inverting the second fraction. Secondly, all terms must be factored completely. Next, common factors in the numerators and denominators may be canceled. Lastly the multiplication operation is performed.

7.3

Student Practice

2. $\dfrac{8}{3x+4}$

4. $\dfrac{2x+11}{(3x-2)(x+9)}$

6. $28(4x+5)$

8. $72a^2b^4c^3$

10. $\dfrac{2xz+4}{xyz}$

12. $\dfrac{11x+22}{4x^2-25}$ or $\dfrac{11(x+2)}{(2x+5)(2x-5)}$

14. $\dfrac{18x+10y}{9x^2-y^2}$ or $\dfrac{2(9x+5y)}{(3x+y)(3x-y)}$

16. $\dfrac{2x+17}{6x+21}$ or $\dfrac{2x+17}{3(2x+7)}$

Extra Practice

1. $(2x-3)(x-4)(x+5)$

2. $\dfrac{3x+5}{x+1}$

3. $\dfrac{2x+5y}{x^2y-y^3}$ or $\dfrac{2x+5y}{y(x+y)(x-y)}$

4. $\dfrac{-x-5}{x^3-7x-6}$ or $-\dfrac{x+5}{(x-3)(x+1)(x+2)}$

Concept Check

Answers may vary. Possible solution:
First factor each denominator completely.
The LCD will be the product containing
each different factor. If a factor occurs
more than once in any one denominator,
the LCD will contain that factor repeated
the greatest number of times that it occurs
in any one denominator.

7.4

Student Practice

2. $\dfrac{4m}{n^2(3m+4)}$

4. $\dfrac{15(m+n)}{mn(3x-5y)}$

6. $\dfrac{x-4}{5(x+2)(x+6)}$

8. $\dfrac{7x-3y}{7}$

10. $\dfrac{16xy-7y^2}{20xy^2-36}$

12. $\dfrac{7x-3y}{7}$

Extra Practice

1. $\dfrac{5a+4b}{2}$

2. $\dfrac{2}{xy}$

3. $\dfrac{2x}{x-35}$

4. $\dfrac{-ab(32x-9y)}{12xy(7b+5a)}$

Concept Check

Answers may vary. Possible solution:
The first step is to find the LCD for the
fractions in the numerator.

$x-3=(x-3)$

$2x-6=2\cdot(x-3)$

$\text{LCD}=2(x-3)$

Next, multiply the first fraction in the
numerator by the 2 to obtain common
denominators in the numerator. Lastly,
add the two fractions in the numerator.

7.5

Student Practice

2. $x=-44$

4. $x=-6$

6. $x=\dfrac{7}{3}$

8. no solution

Extra Practice

1. $x=32$

2. no solution

3. $a=60$

4. $x=-10$

Concept Check

Answers may vary. Possible solution:
To find the LCD first factor each
denominator, then multiply one instance
of each factor.

The math is displayed on the following
page.

$$x^2 - 9 = (x-3)(x+3)$$
$$3x - 9 = 3(x-3)$$
$$2x + 6 = 2(x+3)$$
$$2x^2 - 18 = 2(x-3)(x+3)$$
$$\text{LCD} = 2 \cdot 3 \cdot (x-3)(x+3)$$

7.6

Student Practice

2. 594 miles

4. $17\dfrac{11}{15}$ centimeters

6. Train A traveled 120 kilometers per hour. Train B traveled 105 kilometers per hour.

8. 2 hours and 24 minutes

Extra Practice

1. $x = 38$

2. $133\dfrac{1}{3}$ miles

3. 16 feet

4. 3 hours and 22 minutes

Concept Check

Answers may vary. Possible solution: One of the fractions needs to be inverted in order for the equation to be an accurate statement.

8.1

Student Practice

2. $\dfrac{27}{64}a^3b^{-18}$

4. (a) $x^{12/5}$
   (b) $x^{3/4}$
   (c) $6^{9/13}$

6. (a) $-12x^{9/10}$
   (b) $5x^{7/8}y^{-2/3}$

8. $-8x^{7/6}+24x^{1/2}$

10. (a) 64
    (b) 16

12. $\dfrac{5+x}{x^{3/4}}$

14. $6z\left(3z^{2/3}-4z^{1/3}\right)$

Extra Practice

1. $x^{16/5}$

2. $ab^{4/3}$

3. $\dfrac{1+3y}{y^{2/3}}$

4. $5x\left(2x^{3/4}+5x^{1/8}\right)$

Concept Check
Answers may vary. Possible solution:
Change the exponents to have equal
denominators, then add the
numerators over the common
denominator. This is the combined,
simplified exponent for x.

8.2

Student Practice

2. (a) 4
   (b) 1
   (c) 3.3

4. The domain is all real numbers x,
   where $x\ge 9$.

6. 7

8. (a) $z^{3/5}$
   (b) $a^{8/3}$

10. (a) $\sqrt[3]{a^5b^5}$ or $\sqrt[3]{(ab)^5}$
    (b) $\sqrt[5]{5x}$

12. (a) $\dfrac{1}{4}$
    (b) Not a real number

14. (a) $5|a|$
    (b) $4z^2$
    (c) $3x^2|y|$

Extra Practice

1. 7

2. $\left(4m-3n\right)^{5/8}$

3. $\dfrac{1}{\left(\sqrt[7]{4}\right)^4}$ or $\dfrac{1}{\sqrt[7]{256}}$

4. $7|a^3|b^8$

Concept Check
Answers may vary. Possible solution:
Factor the coefficient completely to
identify the fourth root. Remove the
fourth root from under the radical.
Divide the exponents of the variables
by 4 to remove the variables from
under the radical.

8.3

Student Practice

2. $5\sqrt{2}$

4. $6\sqrt{3}$

6. $-3\sqrt[3]{6}$

8. (a) $4a^2b^2\sqrt{2a}$
   (b) $4yz\sqrt[3]{3x^2yz^2}$

10. $4\sqrt{3a}$

12. $4\sqrt{5}$

14. $5\sqrt{a}-3\sqrt{2a}$

16. $14x^2z\sqrt[3]{5z}$

1. $5x\sqrt{x}$

2. $5b^4\sqrt[4]{ab^3}$

3. $10\sqrt{7}-4\sqrt{5}$

4. $17x\sqrt{3}$

Concept Check

Answers may vary. Possible solution:
Completely factor the coefficient of the
radicand to identify the fourth root.
Divide the exponents of the variables
by 4 to remove the variables from
under the radical. The results are
moved outside the radical. The
remainders stay under the radical.

8.4

Student Practice

2. $-12\sqrt{21z}$

4. $39-10\sqrt{15}$

6. $12+18\sqrt{2}+8\sqrt{3}+12\sqrt{6}$

8. $11-2\sqrt{22a}+2a$

10. (a) $\dfrac{2}{3}$

(b) $6ab^2\sqrt{a}$

12. $\dfrac{5\sqrt{2z}}{8z}$

14. $\dfrac{3\sqrt[3]{a^2}}{a}$

16. $\dfrac{31+7\sqrt{35}}{13}$

Extra Practice

1. $15a\sqrt{ab}$

2. $265+30\sqrt{70}$

3. $\dfrac{3\sqrt{2x}}{5y^2}$

4. $x\sqrt{5}+x\sqrt{3}$

Concept Check

Answers may vary. Possible solution:
To rationalize the denominator, the
numerator and the denominator must be
multiplied by the conjugate of the
denominator. In this case, the conjugate
of the denominator is $3\sqrt{2}+2\sqrt{3}$. The
product will not contain a radical in the
denominator.

8.5

Student Practice

2. $x=6$

4. $x=-\dfrac{1}{3}$ or $x=3$

6. $x=5$

8. $y=3$

Extra Practice

1. $x=3$

2. No solution

3. $x=0$ or $x=4$

4. $x=5$

Concept Check

Answers may vary. Possible solution:
Substitute the found values for x back
into the original equation to test for
validity.

8.6

Student Practice

2. (a) $i\sqrt{41}$

(b) $5i$

4. -18

6. $x=-3,\ y=4\sqrt{2}$

8. $-3+2i$

10. $8-27i$

12. $-24-12i$

14. (a) -1

(b) 1

16. $\dfrac{9-38i}{25}$ or $\dfrac{9}{25}-\dfrac{38}{25}i$

Extra Practice

1. $20-3i\sqrt{2}$

2. $-1+i$

3. $4+7i$

4. $\dfrac{30+12i}{29}$

Concept Check

Answers may vary. Possible solution:
Multiply by the FOIL method.

Replace i^2 with -1. Combine like terms.

8.7

Student Practice

2. 144 feet

4. $y = \dfrac{25}{6}$.

6. 19 lumens

8. $y = \dfrac{175}{6}$

Extra Practice

1. 40 inches

2. $y \approx 3.9$

3. 80 mph

4. $y \approx 4.2$

Concept Check

Answers may vary. Possible solution:

First write the equation as $y = \sqrt{x}k$.

Substitute the known values for x and y and solve for k.

9.1

Student Practice

2. 5 or −5

4. $2\sqrt{6}$ or $-2\sqrt{6}$

6. 4 or −4

8. $5i$ or $-5i$

10. $\dfrac{-8+\sqrt{6}}{3}$ or $\dfrac{-8-\sqrt{6}}{3}$

12. $-2+\sqrt{3}$ or $-2-\sqrt{3}$

14. $\dfrac{2+\sqrt{19}}{5}$ or $\dfrac{2-\sqrt{19}}{5}$

Extra Practice

1. $8i$ or $-8i$

2. $\dfrac{-1+\sqrt{15}}{3}$ or $\dfrac{-1-\sqrt{15}}{3}$

3. −5 or −7

4. $\dfrac{1+i\sqrt{47}}{8}$ or $\dfrac{1-i\sqrt{47}}{8}$

Concept Check

Answers may vary. Possible solution: Divide the coefficient of x (1) by 2 to get $\dfrac{1}{2}$. Since $\left(\dfrac{1}{2}\right)^2=\dfrac{1}{4}$, add $\dfrac{1}{4}$ to both sides of the equation.

9.2

Student Practice

2. $-6\pm\sqrt{29}$

4. $\pm\sqrt{15}$

6. $x\approx 20$ or $x\approx 10$

8. $-1\pm\sqrt{11}$

10. $\dfrac{2\pm i\sqrt{14}}{6}$

12. Complex roots

14. $x^2-7x+12=0$

16. $x^2+45=0$

Extra Practice

1. $\dfrac{-6\pm\sqrt{15}}{3}$

2. $\dfrac{-7\pm\sqrt{89}}{2}$

3. One rational solution

4. $2x^2+7x+3$

Concept Check

Answers may vary. Possible solution: Subtract 3 from both sides to put the equation in standard form. Identify the values of a, b, and c, and find the value of the discriminant, b^2-4ac. If the discriminant is a perfect square, there are two rational solutions; if it is a positive number that is not a perfect square, there are two irrational solutions; if it is zero, there is one rational solution; if it is negative, there are two nonreal complex solutions.

9.3

Student Practice

2. $x=\pm 4,\ x=\pm 5i$

4. $x=3,\ x=\dfrac{\sqrt[3]{18}}{3}$

6. $x=64,\ x=1$

8. $x=625$

10. $x=\dfrac{1}{3},\ x=-\dfrac{1}{4}$

Extra Practice

1. $x=\pm i\sqrt{7},\ x=\pm i\sqrt{3}$

2. $x=0,\ x=-3$

3. $x=-1,\ x=-32$

4. $x=\dfrac{1}{7},\ x=-\dfrac{1}{8}$

Concept Check

Answers may vary. Possible solution: Substitute y for x^4, y^2 for x^8 which yields $y^2-6y=0$. Next factor out the y from the left side of the equation yielding $y(y-6)=0$. Set each term equal to zero and solve for y yielding $y=0$ and $y=6$. Substitute x^4 for y and solve for x yielding $x=0$ and $x=\pm\sqrt[4]{6}$.

9.4
Student Practice

2. $r = \sqrt[3]{\dfrac{3V}{4\pi}}$

4. $x=2y$ or $x=-5y$

6. $x = \dfrac{3y \pm 3\sqrt{y^2-4z}}{4}$

8. (a) $c=\sqrt{a^2+b^2}$
   (b) 25

10. 10 mi, 24 mi, and 26 mi

12. width $=7$ yd; length $=12$ yd

Extra Practice

1. $t = \sqrt{\dfrac{2s}{g}}$

2. $x = \dfrac{7 \pm \sqrt{79-12ay-24y}}{2a+4}$

3. $a=\sqrt{10};\ b=3\sqrt{10}$

4. 25 mph, then 45 mph

Concept Check

Answers may vary. Possible solution: Set one leg's length $=x$, then the other leg's length $=3x$.
Use $c^2=a^2+b^2$ with $c=12$, $a=x$, and $b=3x$. Solve for x to find one leg length then multiply the found value of x by 3 to find the other leg's length.

9.5
Student Practice

2. vertex $=(3.5,-0.25)$;
   y-intercept $=(0,12)$;
   x-intercepts $=(4,0),(3,0)$

4. vertex $=(2,-3)$;
   y-intercept $=(0,1)$;
   x-intercepts $\approx(3.7,0),(0.3,0)$

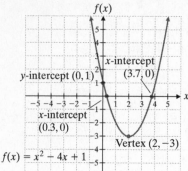

6. vertex $=(-1,-3)$;
   y-intercept $=(0,-6)$;
   No x-intercepts.

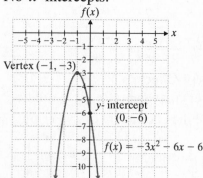

Extra Practice

1. vertex $=(-2.5,-12.25)$;
   y-intercept $=(0,-6)$;
   x-intercepts $=(1,0),(-6,0)$

2. vertex $=(-1.7,-26.45)$;
   y-intercept $=(0,-12)$;
   x-intercepts $=(0.6,0),(-4,0)$

3. $\text{vertex} = \left(\dfrac{1}{3}, \dfrac{2}{3}\right)$ or $\approx (0.3, 0.7)$

$y\text{-intercept} = (0,1)$;

No $x\text{-intercepts}$

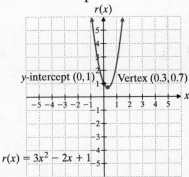

$r(x) = 3x^2 - 2x + 1$

y-intercept $(0,1)$ Vertex $(0.3, 0.7)$

4. $\text{vertex} = (1, 0)$;

$y\text{-intercept} = (0, -1)$;

$x\text{-intercepts} = (1, 0)$

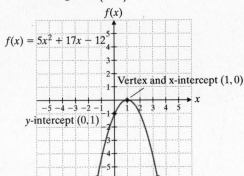

$f(x) = 5x^2 + 17x - 12$

Vertex and x-intercept $(1, 0)$

y-intercept $(0, 1)$

Concept Check

Answers may vary. Possible solution:
The function is in standard form,

$f(x) = ax^2 + bx + c$ with $a = 4$, $b = -9$,

and $c = -5$. The x-coordinate of the vertex

is $x_{\text{vertex}} = \dfrac{-b}{2a} = \dfrac{-(-9)}{2(4)} = \dfrac{9}{8}$. The

y-coordinate of the vertex is

$f(x_{\text{vertex}}) = f\left(\dfrac{9}{8}\right)$.

Student Practice

2.
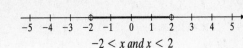
$-2 < x \text{ and } x < 2$

4.
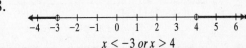
$x < -1 \text{ or } x > 1$

6.
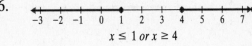
$x \le 1 \text{ or } x \ge 4$

8.
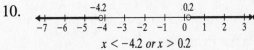
$x < -3 \text{ or } x > 4$

10.
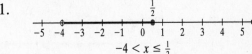
$x < -4.2 \text{ or } x > 0.2$

Extra Practice

1.
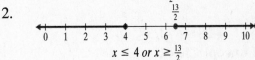
$-4 < x \le \dfrac{1}{2}$

2.
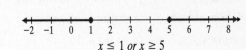
$x \le 4 \text{ or } x \ge \dfrac{13}{2}$

3.
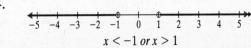
$x \le 1 \text{ or } x \ge 5$

4.
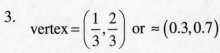
$-1 \le x \le 2.5$

Concept Check

Answers may vary. Possible solution:
The equation $x^2 + 2x + 8 = 0$ does not
have any real solutions. Thus, there are
no boundary points. Also, the quadratic
function $f(x) = x^2 + 2x + 8$ has no

x-intercepts, so the graph lies entirely
above or below the x-axis. This means
that the inequality is either true for all
real numbers, or has no real number
solutions.

9.7

Student Practice

2. $x = 3, \ x = -6$

4. $x = 1, \ x = -2$

6. $x = -2, \ x = -\dfrac{2}{3}$

8.

$-3 \le x \le 3$

10.

$-5 \le x \le 1$

12.

$x \le -4 \text{ or } x \ge 4$

14.

$x \le 1 \text{ or } x \ge 4$

Extra Practice

1. $x = 11, \ x = -20$

2. $m = -2, \ m = \dfrac{2}{3}$

3.

$0 \le x \le 6$

4. $1.28 \le t \le 1.38$

Concept Check

Answers may vary. Possible solution: Since the absolute value is always nonnegative, there is no solution. $\left| 7x + 3 \right|$ cannot be < -4.

10.1

Student Practice

2. $\sqrt{29}$

4. Center at $(1,4)$; radius is $r = 4$

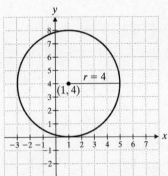

6. $(x-14)^2 + (y+5)^2 = 7$

8. $(x+2)^2 + (y-1)^2 = 9$; Center at $(-2,1)$; radius is $r = 3$

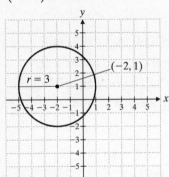

Extra Practice

1. $2\sqrt{5}$

2. $x^2 + \left(y - \dfrac{6}{5}\right)^2 = 13$

3. Center at $(0,0)$; radius at $r = 6$

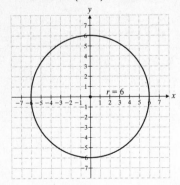

4. $(x+5)^2 + (y-3)^2 = 5$; center at $(-5, 3)$; radius at $r = \sqrt{5}$

Concept Check

Answers may vary. Possible solution: Using the distance formula, let $(x_1, y_1) = (-6,8)$, $(x_2, y_2) = (x,12)$, and $d = 4$. Then, solve for the unknown variable x.

10.2

Student Practice

2. Vertex at $(-1,0)$; axis of symmetry is $x = -1$

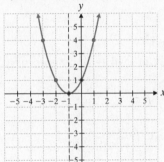

4.

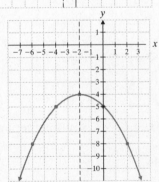

6.

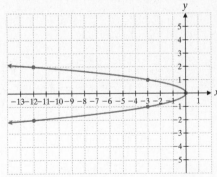

8. $x = (y+3)^2 - 2$

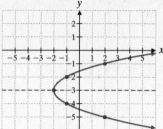

10. $y = -2(x+2)^2 + 4$

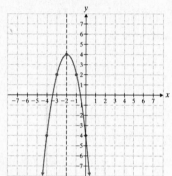

Extra Practice

1. Vertex at $(-4,-3)$; y-intercept at $(0,29)$

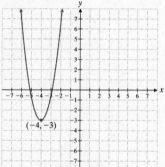

2. Vertex at $\left(\dfrac{3}{2},3\right)$; y-intercept at $(0,-1)$

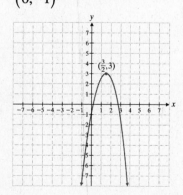

3. Vertex at $(-3,-1)$; x-intercept at $(0,0)$

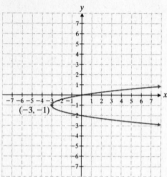

4. $x = -3(y-2)^2 + 18$
(a) Horizontal
(b) Opens left
(c) $(18,2)$

Concept Check
Answers may vary. Possible solution:
$y = ax^2$, vertical, opens up
$y = -ax^2$, vertical, opens down
$x = ay^2$, horizontal, opens right
$x = -ay^2$, horizontal, opens left
Using these rules you can tell which way the parabolas will open.

10.3
Student Practice

2.

4.

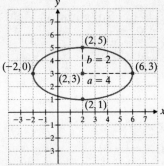

Extra Practice

1.

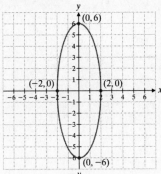

2.

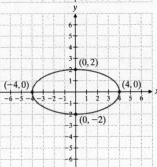

3.

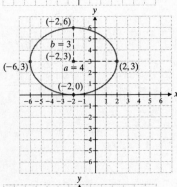

4.

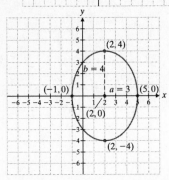

Concept Check

Answers may vary. Possible solution: Add 36 to both sides of the equation, then divide both sides of the equation by 36 to put the equation in standard form. The center of the ellipse is at $(0,0)$ so y-intercepts are $\left(0, \pm\dfrac{b}{2}\right)$ and x-intercepts are $\left(\pm\dfrac{a}{2}, 0\right)$.

10.4

Student Practice

2.

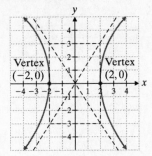

4.

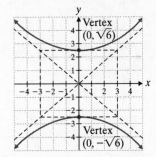

6.

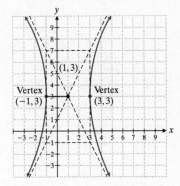

Extra Practice

1.

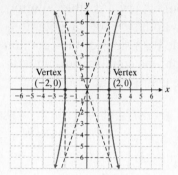

2.

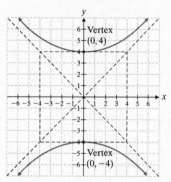

3.

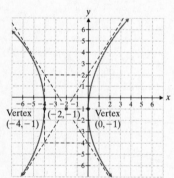

4.

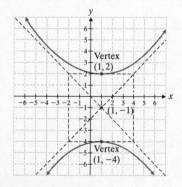

Concept Check

Answers may vary. Possible solution: Divide both sides of the equation by 196 in order to put the equation in standard form. This equation describes a horizontal hyperbola, the asymptote of

which is $y = \dfrac{b}{a}x$. Substitution yields

$$y = \dfrac{7}{2}x.$$

10.5

Student Practice

2. $(0,5),\ (-7,12)$

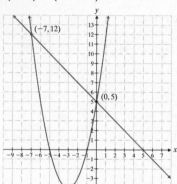

4. $\left(\dfrac{\sqrt{2}}{2}, -\sqrt{2}\right),\ \left(-\dfrac{\sqrt{2}}{2}, \sqrt{2}\right)$

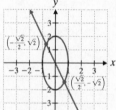

6. $\left(\dfrac{\sqrt{10}}{10}, \dfrac{\sqrt{10}}{10}\right),\ \left(\dfrac{\sqrt{10}}{10}, -\dfrac{\sqrt{10}}{10}\right),$

$\left(-\dfrac{\sqrt{10}}{10}, \dfrac{\sqrt{10}}{10}\right),\ \left(-\dfrac{\sqrt{10}}{10}, -\dfrac{\sqrt{10}}{0}\right)$

Extra Practice

1. $(3,5),\ (-2,0)$

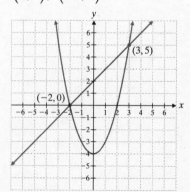

2. $\left(\dfrac{29}{4}, -\dfrac{3}{4}\right)$

3. $(3,4)$, $(3,-4)$, $(-3,4)$, $(-3,-4)$

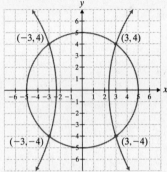

4. $(3,2)$, $(3,-2)$, $(-3,2)$, $(-3,-2)$

Concept Check
Answers may vary. Possible solution:
Use the substitution method. Start by labeling the equations.

$y^2 + 2x^2 = 18$ (1)

$\quad\quad xy = 4$ (2)

Because (2) is a linear equation, and because the y^2 term of (1) has a coefficient of 1, choose to solve (2) for the variable y and substitute the found value for y into (1).

$y = \dfrac{4}{x}$ (2)

$\left(\dfrac{4}{x}\right)^2 + 2x^2 = 18$ (2) and (1)

Solve the resulting equation, in terms of x only, for x.

$x = \pm 1,\ \pm 2\sqrt{2}$

Solve (2) for y, and substitute each of the four found values of x to find corresponding y values.

$(1,4)$, $(-1,-4)$, $\left(2\sqrt{2}, \sqrt{2}\right)$,

$\left(-2\sqrt{2}, -\sqrt{2}\right)$

Check for extraneous answers.

Worksheet Answers Chapter 11

11.1

Student Practice

2. (a) $6b - 7$

   (b) $6b + 29$

   (c) $6b + 22$

4. $\dfrac{-10}{(a+8)(a+6)}$

6. 5

8. (a) 452.16

   (b) $S(e)$

   $= 452.16 + 150.72e + 12.56e^2$

   (c) 498.51 cm^2, the surface area is to large by approximately 46.35 cm^2.

Extra Practice

1. $5a^2 + 2a + 6$

2. $2\sqrt{a^2 + 2}$

3. $\dfrac{5}{2}$

4. $8x + 4h$

Concept Check

Answers may vary. Possible solution:

For the function $k(x) = \sqrt{3x + 1}$

evaluated at $k(2a - 1)$, substitute

$2a - 1$ for x in the function and solve.

$k(2a - 1) = \sqrt{3(2a - 1) + 1} = \sqrt{6a - 2}$

11.2

Student Practice

2. (a) A function

   (b) Not a function

4.
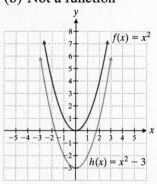

6.
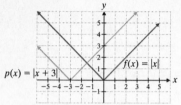

8.
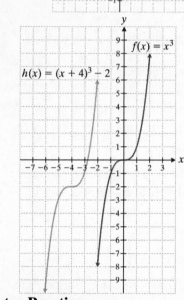

Extra Practice

1. Not a function

2.
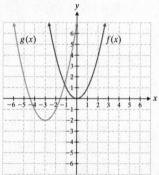

3.
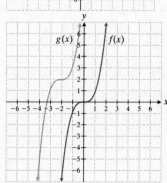

4.

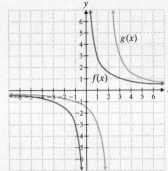

Concept Check
Answers may vary. Possible solution:
If a vertical line passes through more
than one point of the graph of a
relation, the relation is not a function.

11.3
Student Practice
2. (a) $2x^2 + 12x - 6$
    (b) 74
4. (a) $4x^3 - 15x^2 + 41x - 24$
    (b) -198
6. (a) $\dfrac{2x+3}{x-4}$, where $x \neq 4$

    (b) $\dfrac{1}{2x+3}$, where $x \neq -\dfrac{3}{2}$
8. $15x - 17$
10. (a) $\sqrt{4x+5}$
    (b) $4\sqrt{x+6} - 1$
12. (a) $\dfrac{4}{2x-5}$

    (b) -1

Extra Practice
1. (a) $1.6x^3 + 4.6x^2 - 2.7x + 7.6$
    (b) $1.6x^3 - 4.6x^2 - 2.7x - 7.6$
    (c) 84.1
    (d) -33.4
2. (a) $x^3 - 12x^2 + 48x - 64$
    (b) $x - 4$
    (c) -1
    (d) -7
3. $1 - 2x$
4. $\left| -4x - \dfrac{5}{3} \right|$

Concept Check
Answers may vary. Possible solution:
Evaluate both functions for -4, then
subtract the results of $g(-4)$ from the
results of $f(-4)$.

11.4
Student Practice
2. (a) Not one-to-one
    (b) One-to-one
4. (a) Not one-to-one
    (b) One-to-one
6. $Q^{-1} = \{(4,2),(2,6),(7,9),(3,11)\}.$
8. $f^{-1}(x) = \dfrac{x-8}{5}$
10. $f^{-1}(x) = \dfrac{x-75}{3}$
12.

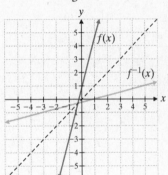

Extra Practice
1. One-to-One
2. $f^{-1}(x) = x + 4$
3. $f^{-1}(x) = \dfrac{5}{3x} + \dfrac{4}{3}$ or $\dfrac{5+4x}{3x}$
4. $g^{-1} = -\dfrac{x+4}{3}$

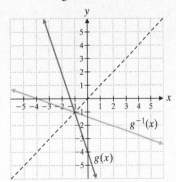

Concept Check
Answers may vary. Possible solution:
To find the inverse of the function

$f(x) = \dfrac{x-5}{3}$, substitute y for $f(x)$.

$y = \dfrac{x-5}{3}$

Interchange x and y.

$x = \dfrac{y-5}{3}$

Solve for y in terms of x.

$x = \dfrac{y-5}{3}$

$3x = y - 5$

$y = 3x + 5$

Replace y with $f^{-1}(x)$.

$f^{-1}(x) = 3x + 5$

12.1

Student Practice

2.

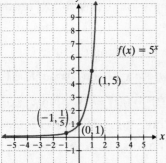

4.

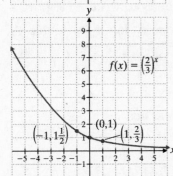

6.

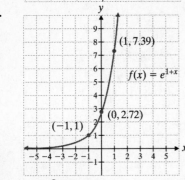

8. $x = 3$
10. $31,990.00
12.   1.34 mg

Extra Practice

1.

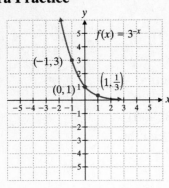

2.

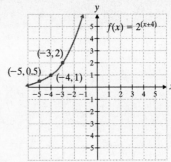

3. $x = -1$
4. $5008.62

Concept Check

Answers may vary. Possible solution:

To solve $4^{-x} = \dfrac{1}{64}$ for x, remember

that $64 = 4^{3}$. Replace the right side of the equation with 4^{-3}, $4^{-x} = 4^{-3}$. This yields the same base on both sides of the equation, and allows the use of the property of exponential functions to simplify, which states that the exponents may be isolated and compared. Thus, $-x = -3$ or $x = 3$.

12.2

Student Practice

2. $-3 = \log_6 \dfrac{1}{216}$

4. $243 = 3^5$

6.
   (a) $x = \dfrac{1}{256}$

   (b) $b = 4$

8.   6

10.

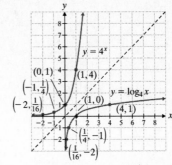

Extra Practice

1. $\log_{10} 0.0001 = -4$

2. $e^{-6} = x$

3. $\dfrac{1}{2}$

4.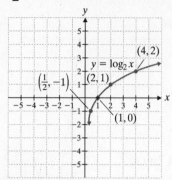

Concept Check

Answers may vary. Possible solution:

To solve $-\dfrac{1}{2} = \log_e x$ for x, write an equivalent exponential equation, $x = e^{-1/2}$.

12.3

Student Practice

2. $\log_5 B + \log_5 C$

4. $\log_2 20xz$

6. $\log_{10} 19 - \log_{10} 5$

8. $\log_b \left(\dfrac{1}{4}\right)$

10. $\log_b \left(\dfrac{y^5 z^3}{w^{1/4}}\right)$

12. $6\log_b w + 5\log_b x - \log_b y$

14. (a) 1
    (b) 0
    (c) $x = 12$

16. $x = \dfrac{49}{125}$

Extra Practice

1. $3\log_7 x + \log_7 y + 2\log_7 z$

2. $\log_a \left(\dfrac{36(6)^{1/3}}{x^5}\right)$

3. $\dfrac{1}{5}$

4. $x = 32$

Concept Check

Answers may vary. Possible solution: To simplify $\log_{10}(0.001)$, start by setting the expression equal to x, $\log_{10}(0.001) = x$. Then, write 0.001 as a power of 10, $0.001 = 10^{-3}$. Now, rewrite the logarithm, $\log_{10} 10^{-3} = x$. Finally, use the logarithm of a number raised to a power property to write $-3\log_{10} 10 = x$, and the property $\log_b b = 1$ to get $-3(1) = x$, or $-3 = x$.

12.4

Student Practice

2. (a) 1.102776615
   (b) 2.102776615

4. $x \approx 27,415,741.72$

6. 0.003132564

8. (a) 1.178654996
   (b) 4.024100909

10. (a) 408.4216105
    (b) 0.376288177

12. 1.159231332

14. -5.295623771

16.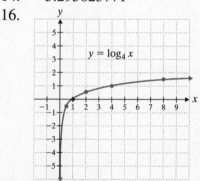

Extra Practice

1. 1.049218023

2. 24.53253019

3. -0.320449841

4. -2.456336474

Concept Check

Answers may vary. Possible solution:
To solve $\ln x = 1.7821$ for x using a scientific calculator, first recall that $\ln x = 1.7821$ is equivalent to $e^{1.7821} = x$.
Then, use the calculator to evaluate $e^{1.7821}$, resulting in $x \approx 5.942322202$.

12.5

Student Practice

2. $x = 4$
4. $x = 5$
6. $x = \dfrac{\log 12}{\log 5}$
8. $x \approx 0.5053$
10. Approximately 17 years
12. Approximately 41 years

Extra Practice

1. $x = 2$
2. $x = 8$
3. $x \approx 6.838$
4. Approximately 10 years

Concept Check

Answers may vary. Possible solution:
To solve $26 = 52e^{3x}$, start by dividing both sides by 52.

$$\frac{1}{2} = e^{3x}$$

Then take the natural logarithm of both sides.

$$\ln\left(\frac{1}{2}\right) = 3x$$

Isolate the x.

$$\frac{\ln\left(\dfrac{1}{2}\right)}{3} = x$$

Approximate the solution with a calculator.

$$x \approx -.0231$$